David W. Taylor
Resistance of Ships and Screw Propulsion

David W. Taylor

Resistance of Ships and Screw Propulsion

ISBN/EAN: 9783954273126
Erscheinungsjahr: 2013
Erscheinungsort: Bremen, Deutschland

© maritimepress in Europäischer Hochschulverlag GmbH & Co. KG, Fahrenheitstr. 1, 28359 Bremen. Alle Rechte beim Verlag und bei den jeweiligen Lizenzgebern.

www.maritimepress.de | office@maritimepress.de

Bei diesem Titel handelt es sich um den Nachdruck eines historischen, lange vergriffenen Buches. Da elektronische Druckvorlagen für diese Titel nicht existieren, musste auf alte Vorlagen zurückgegriffen werden. Hieraus zwangsläufig resultierende Qualitätsverluste bitten wir zu entschuldigen.

RESISTANCE OF SHIPS

AND

SCREW PROPULSION

BY

D. W. TAYLOR

NAVAL CONSTRUCTOR, UNITED STATES NAVY

New York
MACMILLAN AND CO.
AND LONDON
1893

PREFACE.

IN his professional work the writer has often felt the need of a short treatise upon the resistance and propulsion of ships, containing data, formulæ, and tables for use in making estimates.

Such a treatise he has endeavoured to produce.

In handling the subject the papers read at various times before the Institution of Naval Architects by the late Mr. William Froude, and by Mr. R. E. Froude, his son, have been made much use of. This was necessarily the case, since the theories set forth in these papers are now generally accepted, and the experimental results given in them stand alone for accuracy and completeness.

The book contains, however, a good deal of original matter which will, it is hoped, be found of value.

The writer has endeavoured throughout to discuss ships as they are, not floating bodies in general; to set forth methods and deduce results as simple as the nature of the subject will allow, and sufficiently accurate for everyday use.

The proof-sheets have been revised by Professor W. F. Durand of Sibley College, Cornell University. His assistance has been of the greatest value, not only as regards correction of the proof, but also in the discovery and correction of discrepancies in the original text.

D. W. TAYLOR.

U. S. NAVY YARD, MARE ISLAND, CAL,
July 22, 1893.

CONTENTS.

CHAPTER I.

RESISTANCE.

SECT.		PAGE
1.	Preliminary and Definitions .	1
	Symbols used .	2
3.	The Knot	4
4.	Frictionless Submerged Solid	6
5.	Various Kinds of Ships' Resistance	9
6.	Eddy Resistance of Plate .	12
7.	Skin Resistance of Planes .	16
8.	Rankine's Method	20
9.	Froude's Method	25
10.	The Law of Comparison .	28
11.	Froude's Method (Completed)	36
12.	Phenomena of Waves produced by Ships	38
13.	Properties of Trochoidal Waves	43
14.	Deduction of Law of Wave Resistance .	45
15.	Laws of Variation of Wave Resistance .	48
16.	Results of Experiments on Wave Resistance and Approximate Laws	52
17.	Simplification of Wave-Resistance Formula	56
18.	Increase of Resistance in Shallow Water	58
19.	Squat and Change of Trim	59
20.	Formula for Total Resistance .	60

CHAPTER II.

THE PROPELLER.

Sect.		Page
1.	Preliminary and Definitions.	62
2.	Element of Face of Blade treated as a Plane	67
3.	Extension of Formula for Plane to Propeller	76
4.	Values of a, f, and the Characteristics	80
5.	Efficiency and Power of Various Shapes of Blades.	85

CHAPTER III.

MUTUAL REACTIONS BETWEEN PROPELLER AND SHIP.

1.	Action of Propeller attached to Vessel.	95
2.	The Wake.	101
3.	Thrust Deduction	104

CHAPTER IV.

ANALYSIS OF TRIALS AND AVERAGE RESULTS.

1.	Value of Trials	106
2.	Components or Absorbents of the Indicated Horse-Power.	108
3.	Yorktown Trial Analysis.	116
4.	Distribution of Power	133
5.	Coefficients and Constants for Practical Use.	136
6.	Thrust Deduction and Wake Factor	139

CHAPTER V.

THE POWER OF SHIPS.

1.	Preliminary	142
2.	Admiralty Coefficient Method	142
3.	Kirk's Analysis	147
4.	Extended Law of Comparison	151
5.	Standard Curves of Power	155

CONTENTS.

Sect.		Page
6.	Model Tank Method .	180
7.	The Independent Estimate Method	181
8.	Comparison of Methods .	182
9.	Effect of Rough Weather and Foulness .	183

CHAPTER VI.

PROPELLER DESIGN.

1.	Influence of Shape of Section and Variation of Pitch,	186
2.	Standard Blade	194
3.	Standard Slip	198
4.	Design of a Propeller	. 201
5.	Strength of Propeller Blades .	208

TABLES . 222

CHAPTER I.

RESISTANCE.

§ 1. *Preliminary and Definitions.*

If a ship is towed at a steady speed through still water, the pull on the tow-rope must evidently be equal and opposite to the resistance of the ship, which opposes its motion through the water. The resistance of the ship in this simplest case is often called the tow-rope resistance. *[Tow-rope Resistance.]*

When a ship is propelled through still water by her own machinery, there appear in the water various disturbances caused by the propelling machinery which are superposed upon those caused by the ship's hull. Under these circumstances the resistance of the ship generally differs from the tow-rope resistance appropriate to the speed. *[Resistance when Self-Propelled.]*

I shall deal in this chapter with tow-rope resistance, and discuss in a later chapter the effect of the propelling machinery. Used without qualification the term *resistance* will be understood to mean *tow-rope resistance.*

If a ship, instead of being towed through still water, were held by a tow-line in water flowing uniformly and steadily past her, it is evident that, the relative motion being unchanged, the pull on the tow-line for a given speed of the water would be the same as the tow-rope resistance for the same speed of the ship through still water. In discussing resistance I shall consider the ship moving through still water, or fixed in a uniform stream, as is most convenient. *[Ship at Rest in Steady Stream.]*

§ 2. *Symbols Used.*

There will be found below the meanings attached to certain symbols which will be frequently used.

Length. The length of a ship in feet is denoted by L. Since the stem and stern post are not usually vertical, the exact value of L depends upon the level at which the length is taken. For purposes of resistance we need only consider the portion of the hull below water. As a rule it is best to measure L at the level of the water line, making it what is called *length on water line.* Where there are ram spurs or other projections below the water, a reasonable addition should be made to the length on water line to obtain the proper value of L. It would perhaps be a somewhat more nearly exact definition of the symbol L to call it the *mean immersed length of hull.*

Breadth. The greatest breadth of the immersed hull in feet is denoted by the symbol B. This greatest breadth is usually found at the water line, but in some ships the point of maximum breadth is below water.

It should be noted that B is not the molded breadth, or breadth to outside of frames, but should be measured to the outside of the plating or planking where they are flush.

In the case of ships plated on the raised and sunken system, B should be measured to a line $1\frac{1}{2}$ times the thickness of the plating outside the frames; for this line is a fair mean of the irregular contour, being at a distance equal to $\frac{1}{2}$ the thickness of the plating inside the surface of the outer strakes, and the same distance outside the surface of the inner strakes.

Mean Draught. The mean draught in feet at the centre of length is denoted by the symbol H. This draught is exclusive of keel or other projection from the surface. Thus, in the case of a ship with bar keel 12 inches deep, plated on the raised and sunken system, with bottom plating $\frac{3}{4}$ inch thick, and drawing 23 feet amidships to bottom of keel, the value of H would be

$$23' - 12'' + 1\tfrac{1}{2} \times \tfrac{3}{4}'' = 22'.09375.$$

The surface of the hull and its appendages below water, commonly called the *wetted surface or skin*, is expressed in square feet, and denoted by the symbol S. Distinction is and should be made between the wetted surface of the hull proper and that of bar or bilge keels, rudders, shaft tubes, struts, etc. The latter constitute the appendages. *Wetted Surface.*

The displacement in tons is denoted by the symbol D. Of course at a given draught and trim the displacement depends upon the density of the water. It is customary to consider 35 cubic feet of salt and 36 cubic feet of fresh water as weighing one ton. These are convenient round numbers, and sufficiently close approximations. *Displacement.*

There are various ratios or coefficients which may well be defined and explained here, though I shall have occasion to use them but little. *Coefficients.*

If D denote salt-water displacement, then $D \times 35$ denotes the volume displaced in cubic feet.

The ratio $(D \times 35) \div (L \times B \times H)$ is called the *block coefficient*. It is the ratio between the volume of the hull below water and that of a rectangular block of the same extreme dimensions.

The ratio (Area of Midship Section)$\div (B \times H)$ is called the Midship Section Coefficient. It is the ratio between the area of the midship section and that of a rectangle of the same extreme dimensions. It should be remarked that the midship section is the section of maximum area below water, and is not necessarily situated at the centre of length, though it is usually there or thereabouts in ships of the present day.

The ratio (Area of Water Plane)$\div (B \times L)$ is called the Water Plane Coefficient, or Water Line Coefficient.

4 RESISTANCE OF SHIPS. § 3.

The ratio $(D \times 35) \div (\text{Area of Midship Section} \times L)$ is called the Cylindrical Coefficient.

It is the ratio between the volume of the hull below water and that of a cylinder with a section the same as the midship section, and of the same length as the ship.

Resistance. Resistance in pounds is denoted by the symbol R. R used alone will usually denote the total resistance, while used with subscripts, as R_s, R_w, it will denote various components of the total resistance, the subscript indicating the nature of the special resistance denoted.

NOTE.—It will be convenient in other chapters to use R to denote other things than resistance. I shall not hesitate to do so where there is no chance of confusion being caused, as I consider this course preferable to the use of the Greek or German letter R.

Speed. Speed in knots is denoted by V, and speed in feet per second by v.

§ 3. *The Knot.*

Log Line and Chip. The knot, properly speaking, is a unit of speed, and not of length. The expression came into use from the method universally used in the day of sails (and still much employed) for determining the speed of ships at sea. A nearly triangular *log chip* is so weighted and connected to a *log line* that it floats nearly stationary when thrown overboard, and draws the log line off the reel upon which it is wound. Knots are tied in the line at suitable intervals, and by observing the number of knots which pass overboard while a sand-glass is running down, the speed of the ship is determined at once. A common length between knots is 47 feet 3 inches, corresponding to a 28-second sand-glass.

A speed of one knot, then, would mean a speed per hour of

$$\frac{47.25 \times 3600}{28} = 6075 \text{ feet},$$

if the log chip did not move at all. The distance between knots is probably made purposely somewhat less than that corresponding exactly to the nautical mile because of the slight drag of the log chip.

The nautical mile or unit used for measuring distances at sea is not the same in all countries. Since, upon the ocean, latitude is more easily and accurately determined than longitude, we might naturally expect to find the nautical mile and the minute of latitude identical. There is an intimate connection between them. But the arc of a meridian which subtends an angle of one minute at the centre of the earth varies slightly in length from the equator to the poles on account of the fact that the earth is not a perfect sphere. *Nautical Mile in Different Countries.*

Its average length, according to the astronomer Bessel, is 1852.01 metres. Accordingly, the nautical mile used in France, Germany, and Austria is 1852 metres, or 6076.23 feet.

In England the nautical mile corresponding to the "Admiralty knot" is 6080 feet.

In the United States the nautical mile as fixed by the Navy Department is 6080.27 feet, being "equal to the one-sixtieth part of the length of a degree on a great circle of a sphere, the surface of which was considered equal to the surface of the earth."

The "geographical mile" is the length of the arc subtending one minute of longitude at the equator, and is 6086.5 feet long. It appears to be an entirely superfluous unit. *Geographical Mile.*

The "statute mile" used in land measurements is 5280 feet. It is commonly used in navigating river and lake boats, notably on the Great Lakes of America. Such vessels never take astronomical observations to determine their positions, and the statute mile is better suited to their needs. *Statute Mile.*

While, as stated above, the word *knot* is properly restricted to denote a unit of speed, the expression *nautical mile* or *sea mile* is rather clumsy and tends to produce confusion. Accordingly, there is observed a growing tendency to use the word *knot* in the sense of nautical mile. *Usage of Word Knot.*

This usage is convenient, and though strenuously opposed as a solecism by the grammatical purist and the amateur sailor, it appears probable that in time it will prevail.

§ 4. *Frictionless Submerged Solid.*

Solid in Perfect Fluid.

It is advisable to lead up to the somewhat complex practical case of the resistance of a ship by a short discussion of a simple ideal case of no resistance.

Suppose we have held deeply submerged, in a *perfect* and incompressible fluid, a small, smooth, or frictionless solid with sharp head and tail, and of fair and easy lines all over.

For our purposes it suffices to define a perfect fluid as one destitute of viscosity.

Imagine the whole body of fluid to be flowing steadily and horizontally in one direction, and the solid to grow gradually from nothing, so to speak, being kept at rest by a force which does not interfere with the motion of the fluid. We know from the mechanics of fluids that after the solid has attained its full growth the surrounding fluid will continue to flow steadily past it. The direction and magnitude of the velocity at each point of the fluid will not change with time, though they will differ from point to point.

Stream Lines.

The paths followed by the particles of fluid under the above conditions are called the "stream lines" past the solid. Their shape depends upon the shape of the solid, and does not change with change of speed of flow of the fluid.

Plane Stream Lines past Cylindrical Solid.

Figure 1 shows some stream lines past a cylindrical solid, a quarter-section of which is shown in the figure. It is supposed to be placed vertically in the fluid, and while of small dimensions relative to the fluid to have its ends at such a distance from the section shown in Figure 1, that the flow past this section goes on as if the solid were indefinitely long. Figure 1 shows the stream lines in one quadrant only, but they are symmetrical in the remaining three.

§ 4. FRICTIONLESS SUBMERGED SOLID.

At every point along a stream line what is called the steady motion formula holds good, or, **Steady Motion Formula.**

$$\frac{p}{w} + \frac{v^2}{2g} + z = h.$$

In this formula

$p =$ pressure per unit area;
$w =$ weight of unit volume of the fluid;
$v =$ velocity of flow;
$g =$ acceleration due to gravity;
$z =$ height above a fixed level;
$h =$ a constant for each stream line, being *called the head.*

If, then, the motion is horizontal, it is evident that along a stream line increase of velocity must accompany decrease of pressure, and *vice versa.* **Changes of Velocity and Pressure along a Stream Line.**

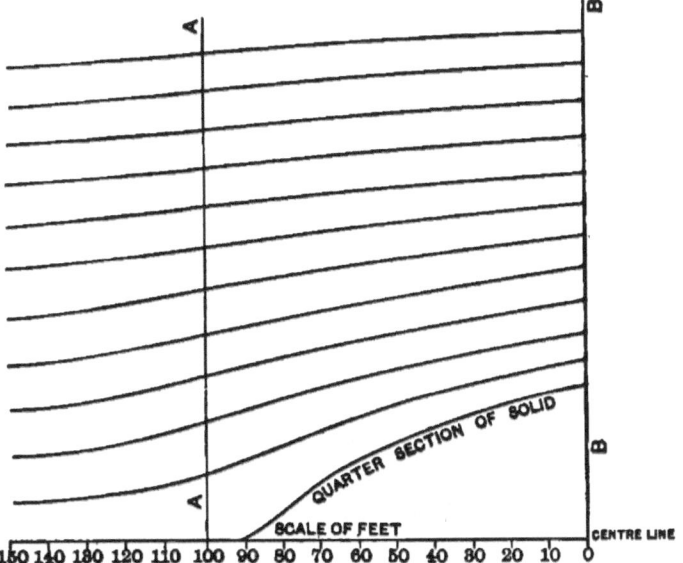

Fig. 1.— STREAM LINES PAST A CYLINDRICAL SOLID.

Consider a stream line passing close to our submerged solid. At a great distance ahead the pressure and velocity are constant, having what may be called their undisturbed or *normal* values. As the solid is approached, the pressure increases and the velocity falls off. We reach a point, however, somewhere near the head of the solid, where the velocity begins to increase again.

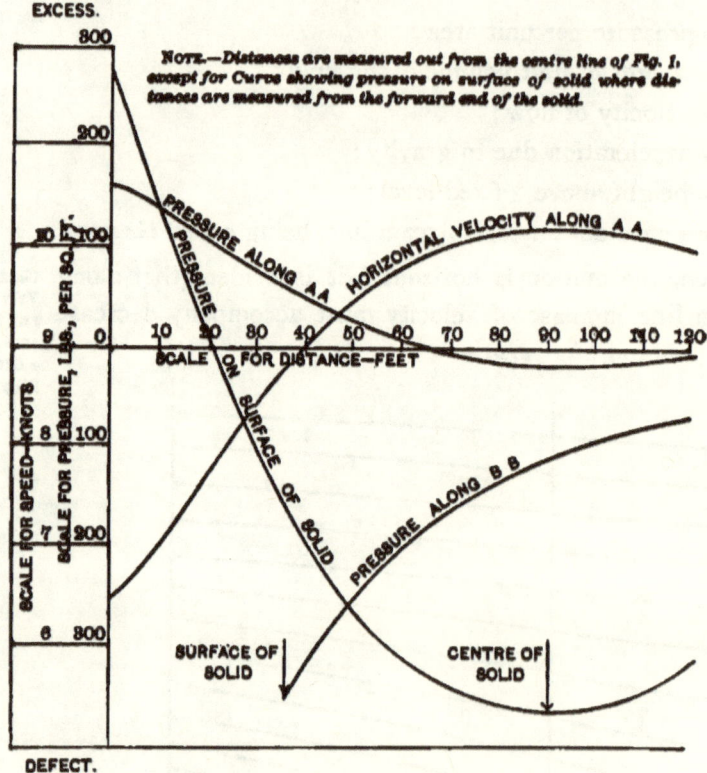

Fig. 2.—CURVES OF PRESSURE AND VELOCITY AT SECTIONS SHOWN IN FIG. 1.

It continues to increase until abreast the centre of the solid we have a velocity above the normal accompanied by a pressure below the normal. This is necessarily the case; for abreast the centre the stream lines are parallel to their directions at a great distance ahead (where the pressure and velocity are normal), but the area available for flow is diminished by the sectional area of the solid.

§ 5. VARIOUS KINDS OF SHIPS' RESISTANCE.

After passing the centre of the solid the changes of velocity just described are repeated in reverse order, until at a great distance astern the flow again becomes normal.

It appears, then, that at what may be called the bow and stern of the submerged solid the pressure in the fluid is greater than if the solid were not there, while amidships it is less.

Figure 2 shows the changes of velocity and pressure set up at the sections indicated in Figure 1. The numerical values were obtained by supposing the solid to be 180 feet long, immersed in a fluid weighing 64.32 pounds per cubic foot, and flowing at the rate of 10 knots.

The densest salt water weighs nearly 64.32 pounds per cubic foot, and that number was adopted for convenience because it is twice 32.16, the value of g, the acceleration due to gravity.

Figure 2 also contains a curve showing the pressure on the surface of the solid at each point, or rather the change in pressure from its value for fluid at rest. It is evident that the net result of the changes in pressure is nil. Since by supposition the fluid is frictionless, and since it exerts no resultant pressure ahead or astern on the solid, it is evident that in the ideal case supposed the solid offers no resistance. *Net Resistance of Solid is Nothing.*

We know very well that ships offer a great deal of resistance, and in the next section I shall point out how the various resistances are brought about as we proceed from the ideal solid to the actual ship.

NOTE.—The theory of stream lines is beyond the scope of this work, so I have contented myself in the above with simple statements of facts. In any modern treatise on hydrodynamics the reader will find the steady motion formula demonstrated, and the theory of stream lines set forth.

§ 5. *Various Kinds of Ships' Resistance.*

I propose in this section to enumerate the points of difference between the ideal solid in a frictionless fluid and a ship in water, pointing out the various kinds of resistance arising from these differences.

Limits of Stream-Line Motion.

We have seen that for a stream line $\frac{p}{w}+\frac{v^2}{2g}=$ a constant $=\frac{p_0}{w}+\frac{v_0^2}{2g}$, where p_0 and v_0 are the normal pressure and velocity, or the pressure and velocity at a great distance.

Then
$$p-p_0=\frac{w}{2g}(v_0^2-v^2).$$

Suppose $v=nv_0$, n is a quantity depending entirely upon the location of the point at which the velocity is v.

Then
$$p=p_0-\frac{w}{2g}\times v_0^2(n^2-1).$$

So if p_0 remains the same while v_0 is increased, we shall finally (n being greater than unity) reach a negative value of p. This is impossible in a fluid. Instead of flowing in such a way as to show a negative pressure, *i.e.* a tension, the fluid will break away from the solid, and flow along a new surface which, though wavering and mutable, marks a line of separation between fluid which is flowing in stream lines, and the *broken, dead,* or *eddying* fluid which surrounds the rear of the solid.

It would appear at first sight that since the minimum stream-line pressure is abreast the centre of the solid, the eddies should form there. In a perfect fluid, *flowing in exact stream lines*, it is probable that eddying would first appear amidships; but as soon as it appeared, the whole of the stream-line motion aft of the centre, in the vicinity of the solid, would be interfered with, and the fluid would break away from the stern of the solid and close in again amidships, because the fact of following the new surface would increase the pressure amidships.

Eddy Resistance.

The pressure in the eddying fluid aft is less than the natural stream-line pressures would be, and this defect of pressure causes resistance.

This kind of resistance is called Eddy Resistance. I shall denote it by the symbol R_e.

Skin Resistance.

The perfect fluid in which I supposed the ideal solid to be immersed is a mathematical concept and does not exist in nature. Water is not a perfect fluid, and we do not in prac-

§ 5. VARIOUS KINDS OF SHIPS' RESISTANCE.

tice meet with smooth surfaces over which it glides freely. Such surfaces as ships' bottoms exercise a certain drag upon water flowing past, thus causing resistance.

This kind of resistance is conveniently termed Skin Resistance. I shall denote it by the symbol R_r.

There is a third highly important point of difference between the ideal solid and a ship. A ship floats upon the surface where the pressure is constant. So stream-line motion with its constant change of pressure along the stream line is at once interfered with. There is, of course, excess pressure developed at the bow. For perfect stream-line motion this excess pressure would be absorbed in changing the stream-line velocity. In the case of the ship, part of the excess pressure is absorbed in changes of velocity; but part leaks away, so to speak, and is used up in causing elevation of the surface. *Wave Resistance.*

But elevations and depressions of the surface of water — in a word, waves — will not stay by the ship. They spread out into undisturbed water and carry with them the energy required to produce them. Thus there is a constant drain of energy from the vicinity of the ship which must be made good by a resistance acting upon the water.

A strictly accurate name for this kind of resistance would be "Wave Making Resistance," but I shall use the term "Wave Resistance," as being shorter, simpler, and sufficiently definite. I shall denote this kind of resistance by the symbol R_w.

The viscosity of water, which largely affects skin resistance, also causes resistance directly, as for instance by opposing the change of shape which a small cube of water would undergo during stream motion in a curved line past the ship's hull. *Viscosity.*

Resistance due to this cause is very small, and may be neglected except for small models at low speeds.

We shall see later that, having determined the resistance of a model, the resistance of the full-sized ship may be closely estimated, and that for a given speed of ship the smaller the model, the smaller the speed at which its resistance is needed.

The disproportionate viscosity resistance of small models at low speeds renders their use undesirable. This is about the only practical conclusion to be drawn from our somewhat limited knowledge of viscosity resistance.

Air Resistance. The above-water portions of a ship may be said to be immersed in the air, which of course offers some resistance to the motion of the ship through it.

This resistance is so small, comparatively, that it may be neglected except for a type that is obsolete, — or nearly so, — the full rigged steamer.

For a rough approximation it may be said that the resistance of the air in pounds may be expressed by $.005 \, AV^2$, where A is the area in square feet of the upper works, rigging, etc., opposing the air, and V is the speed in knots of the air past the ship.

There are then three principal components, which need to be considered in detail. We may say, Total Resistance = Eddy Resistance + Skin Resistance + Wave Resistance, or, in symbols,

$$R = R + R + R_w.$$

§ 6. *Eddy Resistance of Plate.*

Before taking up in detail the components of a ship's total resistance I shall consider briefly the eddy resistance and skin resistance of a thin, flat, submerged plate. As it is convenient to discuss them separately, I shall, in dealing with the eddy resistance, take the plate as frictionless and fully submerged.

Head and Tail Resistance. In this condition let a denote the angle between the face of the plate and the direction of undisturbed flow of the water. Then (Fig. 3) we shall have in front of the plate nearly exact stream-line motion, while in the rear will be broken water and eddies. There will then be an excess of pressure on the front face causing a "*head resistance*," and a defect of pressure on the rear face causing a "*tail resistance.*" Suppose now we fit behind the plate a frictionless solid such that, as the water comes around the edges of the plate, it flows off over the smooth solid with perfect stream motion.

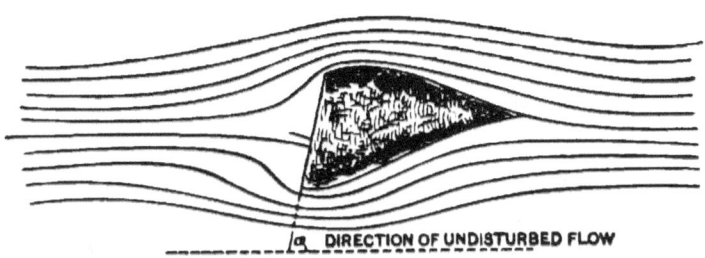

Fig. 3. — THIN PLATE WITH EDDIES BEHIND.

The introduction of the solid would evidently make little or no change in the flow in front of the plate, or in the head resistance.

By the aid of the above artifice Lord Rayleigh has deduced the following formula for the total normal pressure on the front face of the plate: *Rayleigh's Formula.*

$$P_n' = \frac{2\pi \sin a}{4 + \pi \sin a} \frac{w}{2g} A v^2.$$

In the above P_n' = the normal pressure in pounds, w = weight in pounds of a cubic foot of water, g = acceleration due to gravity in feet per second, A is the area of the plane in square feet, v is the speed of advance of the plane in feet per second, and a is the inclination of the face of the plane to the direction of advance.

The tail resistance cannot be treated mathematically. It is commonly assumed that it follows the same law of variation as the head resistance — increasing as the square of the speed. This seems an allowable assumption for the majority of practical cases. It is evident, though, that if the speed be *Tail Resistance.*

so great that the rear pressure is zero, then a further increase of speed will not be accompanied by less pressure or more tail resistance.

Total Eddy Resistance. Since we cannot treat the whole of the eddy resistance of a plate mathematically, it is necessary to have recourse to experiments — deducing from them more or less approximate formulæ to cover the ground. It should be noted that in making experiments eddy resistance will be mixed up with skin resistance, but the latter is comparatively small except at very small angles.

Jœssel's Formula. Reliable experiments upon the total resistance of planes were made by M. Jœssel in 1873. They were made in the river Loire, at Indret, near Nantes. Unfortunately the maximum speed of current in which experiments were made was only about 2½ knots.

M. Jœssel aimed to determine —

1. The total normal pressure on a plane inclined at any angle.

2. The distance of the centre of pressure, or line about which the plane would pivot, from the leading edge.

Jœssel's conclusions as to normal pressure are expressed by the following formula, in which P_n denotes the total normal pressure (covering both head and tail resistance) in pounds, and the other symbols have the same meaning as in Rayleigh's formula:

$$P_n = 1.622 \frac{\sin a}{.39 + .61 \sin a} \cdot \frac{w}{2g} \cdot Av^2.$$

As to the centre of pressure Jœssel concluded that if l denote the breadth of the plane in the direction of motion, and x the distance of the centre of pressure from the leading edge,

$$x = (.195 + .305 \sin a) \, l.$$

Jœssel's total-pressure formula, while doubtless fairly accurate for moderately large angles of inclination, does not hold

§ 6. EDDY RESISTANCE OF PLATE. 15

for small angles of 10° or less when the plane is advancing nearly parallel to itself.

Special and reliable experiments were made upon plates at small inclinations by Mr. William Froude. He gives the following as a fair expression for the normal pressure in fresh water upon planes about 3 feet wide: *Froude's Formula for Small Values of α.*

$$P_n = 1.7 \sin \alpha \, A v^2.$$

Here the symbols have the same meaning as before.

For small angles Rayleigh's formula for planes in fresh water may be written, *Comparison of Formulæ.*

$$P_n = 1.534 \sin \alpha A v^2,$$

while Jœssel's formula becomes,

$$P_n = 4.041 \sin \alpha A v^2.$$

Evidently, then, Froude's formula takes but little account of eddying behind the plate.

Certain conclusions deduced by Lord Rayleigh as to the manner of flow past a plate are of interest and may be stated here. *Point of Division of Stream Past Plane.*

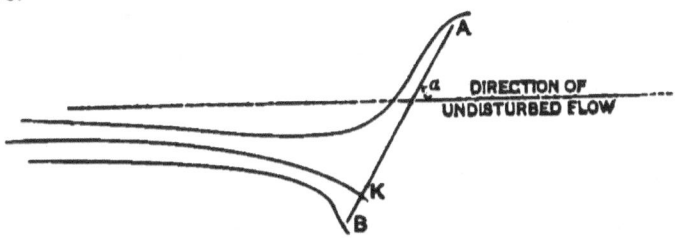

Fig. 4.—FLOW PAST INCLINED PLANE.

Referring to Figure 4, suppose AB a section of a plane at an angle α with the direction of undisturbed flow. The stream will divide at some point K, part flowing around by A, and the remainder by B. The point K is so situated that

$$\frac{AK}{AB} = \frac{2 + 4\cos\alpha - 2\cos^2\alpha + (\pi - \alpha)\sin\alpha}{4 + \pi \sin\alpha}.$$

Friction on Front Face.

Between K and A the water next the plane is flowing with variable velocity toward A, and between K and B is flowing in the opposite direction.

The resultant friction on the front face is of course less than if the flow were all in one direction with a uniform velocity equal to the normal speed of the stream; *i.e.* if the plane had no inclination.

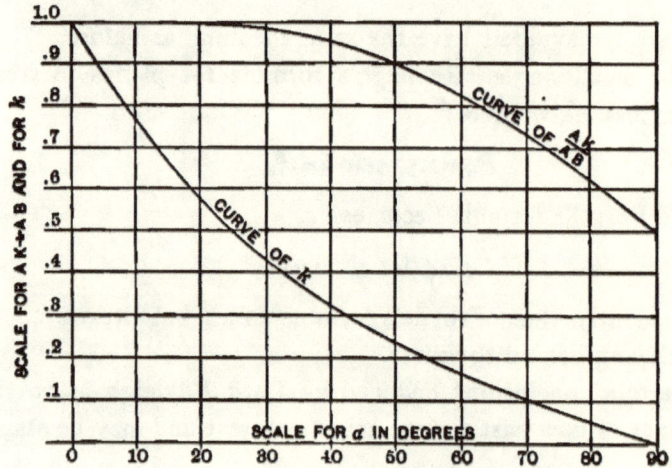

Fig. 5. — DIVISION OF STREAM PAST AND FRICTION ON FACE OF INCLINED PLANE.

Cotterill's Expression for Friction.

Let k denote the ratio between the friction when inclined and when there is no inclination. Professor Cotterill has shown — using Rayleigh's results — that

$$k = \frac{4\cos a - (\pi - 2a)\sin a}{4 + \pi \sin a}.$$

Figure 5 shows graphically the value of $\frac{AK}{AB}$ and of k, for values of a from 0° to 90°.

§ 7. *Skin Resistance of Planes.*

Reliance on Experiment Necessary.

Mathematical analysis has not as yet been successfully applied to skin resistance. For its determination we must rely upon results and formulæ obtained by experiment.

§ 7. SKIN RESISTANCE OF PLANES.

Reliable and accurate experiments upon the skin resistance of planes were made by the late Mr. William Froude, about 1870.

Boards $\frac{3}{16}$ of an inch thick, 19 inches deep, of various lengths (the maximum being 50 feet) and coated with various substances, were towed lengthwise in a tank of fresh water 300 feet long — their resistances at various speeds being carefully measured.

Froude has summarised the results of these experiments in the following table, which is worthy of attentive study.

Froude's Experiments.

Froude's Conclusions.

TABLE I.

Results of Froude's Experiments upon Skin Friction.

Nature of Surface.	Length of Surface or Distance from Cutwater.											
	2 feet.			8 feet.			20 feet.			50 feet.		
	A	B	C	A	B	C	A	B	C	A	B	C
Varnish	2.00	.41	.390	1.85	.325	.264	1.85	.278	.240	1.83	.250	.226
Paraffin	1.95	.38	.370	1.94	.314	.260	1.93	.271	.237			
Tinfoil	2.16	.30	.295	1.99	.278	.263	1.90	.262	.244	1.83	.246	.232
Calico	1.93	.87	.725	1.92	.626	.504	1.89	.531	.447	1.87	.474	.423
Fine sand	2.00	.81	.690	2.00	.583	.450	2.00	.480	.384	2.06	.405	.337
Medium sand	2.00	.90	.730	2.00	.625	.488	2.00	.534	.465	2.00	.488	.456
Coarse sand	2.00	1.10	.880	2.00	.714	.520	2.00	.588	.490			

In the above, for each length stated in the heading —

Column *A* gives the power of the speed according to which the resistance varies.

Column *B* gives the mean resistance in pounds per square foot of the whole surface for a speed of 600 feet per minute.

Column *C* gives the resistance in pounds, at the same speed, of a square foot at the distance abaft the cutwater stated in the heading.

Formula for Skin Resistance.

From the above and other experiments it appears that the skin resistance of a given plane can be expressed very closely by the formula,

$$R_s = f S V^n,$$

where R_s = the Skin Resistance, in pounds;

f = a constant, the "coefficient of friction" for the plane in question;

S = the total frictional area in square feet;

V = the speed in knots;

n = a constant index.

Table II. below gives values of f and n corresponding to Froude's results in Table I.

TABLE II.

Coefficient and Index of Friction (from Froude's Results).

Nature of Surface.	Length of Surface.							
	2 feet.		8 feet.		20 feet.		50 feet.	
	n	f	n	f	n	f	n	f
Varnish .	2.00	.0117	1.85	.0121	1.85	.0104	1.83	.0097
Paraffin .	1.95	.0119	1.94	.0100	1.93	.0088		
Tinfoil	2.16	.0064	1.99	.0081	1.90	.0089	1.83	.0095
Calico .	1.93	.0281	1.92	.0206	1.89	.0184	1.87	.0170
Fine sand	2.00	.0231	2.00	.0166	2.00	.0137	2.06	.0104
Medium sand .	2.00	.0257	2.00	.0178	2.00	.0152	2.00	.0139
Coarse sand	2.00	.0314	2.00	.0204	2.00	.0168		

The Coefficient of Friction.

The value of the coefficient f depends upon a variety of circumstances.

1. The coefficient f depends upon the nature of the surface. This is a matter of course.

2. The coefficient f depends upon the nature of the fluid in which the surface is immersed. For small variations, as for

§ 7. SKIN RESISTANCE OF PLANES. 19

change from fresh to salt water, f varies directly as the density.

3. The coefficient f depends upon the length of the frictional surface, falling off as the length increases. This is because the rear portion of a plane is in contact with water which has a forward motion imparted by the friction of the front portion of the plane.

It is evident from Table I. that the friction per square foot is diminishing very slowly as we go aft at a distance of 50 feet from the cutwater. This enables us to estimate with fair accuracy the friction of planes much longer than 50 feet. But these days of 600-foot ships' experiments on longer planes are much needed.

4. The coefficient f changes slightly with the temperature, decreasing as the temperature increases. This change is very small for the changes of temperature met with in practice, and for our purposes it may be neglected.

5. The coefficient f is entirely independent of the pressure of the water and the depth below the surface.

It is well established that for smooth surfaces of length greater than a few feet, the frictional index n should be less than 2. Before Froude's experiments it was commonly assumed that the friction always varied as the square of the speed, but the lower power is now accepted for surfaces like those of ships. *The Index of Friction.*

Froude adopted the lower power because it agreed very closely with his experimental results and appears to have been entirely justified in so doing. For short surfaces Froude found, however, that the law of the square applied, or nearly so, as will appear from an inspection of Table I.

The diminution of the frictional index with increase of length is, no doubt, due to the same cause as the diminution of the coefficient of friction. Exactly how it comes to be brought about does not appear.

Additional and more extended experiments on the subject are much needed.

Difference between Law of 1.83 and Law of Square.

The value 1.83 of the frictional index, appropriate to long, smooth surfaces, gives results differing from those obtained by supposing $n = 2$ much more than might be thought at first sight. This difference, too, becomes proportionately greater as the speed increases. These facts are illustrated by the table below.

Speed in knots, V.	1	5	10	15	20	25
V^2	1	25	100	225	400	625
$V^{1.83}$	1	19.02	67.61	141.99	240.37	361.60
$V^{1.83} \div V^2$	1	.761	.676	.631	.609	.579

§ 8. *Rankine's Method.*

Augmented Surface Method.

We are now in a position to discuss the resistance of a ship. I propose in this section to give a demonstration of the Augmented Surface Method, which originated with Rankine, and to point out its weak points. We now know that Rankine's Method is of little practical value, because some of the fundamental assumptions upon which it is based are untenable. The method is, however, a beautiful and most instructive example of the application of mathematical principles to our subject, and is worthy of attentive study.

Assumptions Made.

Rankine assumed that in well-formed and properly proportioned ships wave and eddy resistance are negligible, the total resistance differing but little from the skin resistance.

This latter he assumed to vary as the square of the speed. Furthermore, Rankine assumed that water moved past a ship with perfect stream motion. This being the case, velocity of gliding over the skin would be in some places greater, in others less, than the speed of the ship.

Relation between Augmented and Wetted Surface.

Rankine then proceeded to deduce an "augmented surface" bearing such a relation to the actual wetted surface of a ship that the loss of energy through skin resistance in a stream flowing uniformly with the speed of the ship past the plane augmented surface would be the same as the energy absorbed by the skin resistance of the actual wetted surface.

Then, from what has gone before, if R_a denote what may be called the augmented surface resistance, S_a the augmented surface itself, and V the speed in knots, $R_a = f S_a V^2$, f being a suitable frictional coefficient. *Formula.*

Rankine completed his method by giving to f a value deduced from experience of ships in existence at that time — some thirty years ago.

The augmented surface method may be deduced as follows: *Friction of an Element.*

Referring to Figure 6, let dS be an element of the wetted surface of a vessel in a stream which has a normal speed of

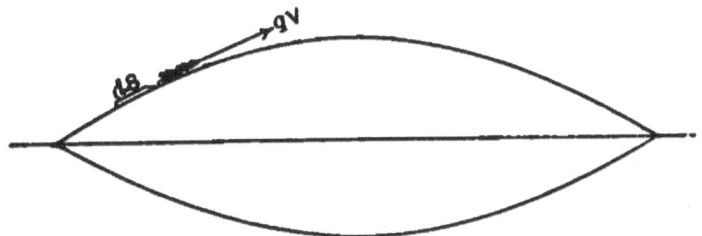

Fig. 6. — FOR AUGMENTED SURFACE METHOD.

flow denoted by V. The speed of flow over the element dS, while in general differing from V, will bear a constant ratio to it whatever its value. Then it may be denoted by qV, where q is a geometrical coefficient, depending upon the shape of hull, position of dS, etc., but not upon V. Then if dR_a' be an elementary resistance due to the friction of dS (being exerted of course parallel to dS), we have on Rankine's assumptions $dR' = f dS (qV)^2$.

Also the corresponding loss of energy per unit time
$= dR_a' \times qV = f(dS)(qV)^3$.

Total Friction. Whence the total loss of energy per unit time $= \int f q^3 V^3 dS$, the integration, extending over the whole of the wetted surface.

But the total loss of energy per unit time due to the resistance = the work done in overcoming the resistance $= R_a V$.

Whence, $$R_a V = \int f q^3 V^3 dS,$$

or $$R_a = f V^2 \int q^3 dS = f V^2 S_a = f V^2 S q_a,$$

where S^a is the augmented surface, S is the actual wetted surface, and q_a is the "coefficient of augmentation."

Now a ship is not of the form of a geometrical solid, so we cannot integrate over its surface. Also, we have no expression for the value of q at every point. We need then a reliable method of approximation.

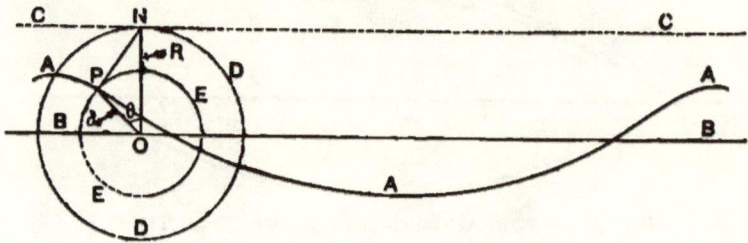

Fig. 7. — SECTION OF TROCHOID.

Trochoidal Ribbon. I wish now to call attention to Figure 7, which refers to a trochoidal ribbon of unit breadth, a section of which is the trochoid AAA. NDD is the generating circle rolling on CC, and P is the generating point.

Denote ON by R, OP by r, and the angle NOP by θ.

Water is supposed to glide over AAA, the normal velocity of the stream being parallel to BB, and equal to the velocity along BB of the centre O when the generating circle rolls steadily on CC with angular velocity $= \omega$.

§ 8. RANKINE'S METHOD.

Under these circumstances it is reasonably assumed that the velocity of gliding at P is the same as the velocity of the generating point P along AA.

This determines the value of the coefficient q at any point. For N is the instantaneous centre of the rolling circle, so that velocity of $P = NP \times \omega$. But $V = R\omega$, or $\omega = \dfrac{V}{R}$.

Whence,
$$\text{velocity of } P = \frac{NP}{R} \times V = qV,$$

or
$$q = \frac{NP}{R}.$$

Also, if $d\theta$ denote the small angle rolled through in small time dt,
$$dS = NP \cdot \omega dt = NP \cdot d\theta.$$

We have seen above that
$$S_a = \int q^3 dS.$$

Substituting for q and dS their values in terms of NP and θ, we have,
$$S_a = \int \left(\frac{NP}{R}\right)^3 NP d\theta.$$

For the augmented surface from crest to crest of the trochoid the limits are o and 2π, whence
$$S_a = \int_0^{2\pi} \left(\frac{NP}{R}\right)^3 NP \cdot d\theta = \frac{1}{R^3} \int_0^{2\pi} \overline{NP}^4 d\theta.$$

Now $\overline{NP}^2 = \overline{ON}^2 + \overline{OP}^2 - 2\,ON \cdot OP \cos NOP$
$$= R^2 + r^2 - 2Rr \cos \theta.$$

Whence,
$$S_a = \frac{1}{R^3} \int_0^{2\pi} (R^2 + r^2 - 2Rr \cos \theta)^2 d\theta.$$

Upon integrating out between the limits o and 2π, we have

$$S_a = 2\pi R\left[1 + 4\left(\frac{r}{R}\right)^2 + \left(\frac{r}{R}\right)^4\right].$$

Now NP is a normal to the trochoid and the inclination of the tangent at P to BB (the path of the centre O) is equal to the angle ONP, and is called the angle of obliquity at the point P, or simply the "obliquity."

For the maximum obliquity, denoted by a, OPN must evidently be a right angle. In that case

$$\sin a = \frac{OP}{ON} = \frac{r}{R}.$$

Whence the augmented surface formula above may be written $S_a = 2\pi R(1 + 4\sin^2 a + \sin^4 a)$, where the expression in brackets is called the coefficient of augmentation, q_a.

It should be noted that this coefficient is applied not to the actual curved length of the trochoid, but to $2\pi R$, the distance in a straight line between successive crests.

Application to Ship.

Rankine made application of the result above as follows:

He divided the wetted surface of a ship by a number of equidistant horizontal planes and assumed that he thus obtained on each side a number of trochoidal ribbons.

Measuring the maximum inclination toward bow and stern of each ribbon, he obtained a number of values of $\sin a$.

Then by Simpson's Rules he obtained the mean value of $1 + \sin^2 a + \sin^4 a$ this mean value being q_a, the "coefficient of augmentation." The mean immersed girth $(= G)$ was then determined by measurement of numerous sections and the application of Simpson's Rules.

Then augmented surface $= S_a = q_a G \times L$, and the augmented surface resistance $= R_a = f S_a V^2$.

For resistance in pounds, surface in feet, and speed in knots, Rankine gave f the value .01, whence

$$R_a = .01\, S_a V^2.$$

There are two principal defects in the Augmented Surface Method outlined above. *Errors of Augmented Surface Method.*

First and foremost, wave and eddy resistance are by no means negligible in comparison with skin resistance, except for large ships at very moderate speeds.

In the second place, water does not flow past the hull of a ship with perfect stream motion. The friction of the ship's surface and the changes of level of the water surface both interfere with the kind of stream motion upon which the Augmented Surface Method is based. It is impossible to determine exactly how much the stream-line velocities assumed by Rankine would be interfered with by the disturbing elements above noted, but they must seriously diminish the true value of the coefficient of augmentation, even if they do not reduce it to unity.

For fast ships, where the wave resistance constitutes a large portion of the total, the Augmented Surface Method is clearly of no use. *Augmented Surface Method Obsolete.*

For the 9 or 10 knot ships for which it might be used there are other much simpler approximate methods which exceed it in accuracy.

Hence we find that this method, though taken up enthusiastically when first brought forward by Rankine, has been almost wholly discarded at the present time.

§ 9. *Froude's Method.*

The most accurate method at present known for approximating to the total resistance of a ship is that developed by the late Mr. William Froude. In this method the skin resistance is calculated directly, and the combined wave and eddy resistance deduced from the results of towing experiments upon a model of the ship. The sum of the wave and eddy resistance is called the Residuary Resistance and denoted in my notation by R_r. *Separation of Elements of Resistance.*

Skin Resistance.

Froude concluded after many experiments that the skin resistance of a ship is practically the same as that of a plane surface of the same nature, area, and length in the direction of motion as the curved wetted surface of the ship. This conclusion was shared by Dr. Tideman, the well-known Dutch naval architect, who had also made numerous experiments upon the subject. We have seen in discussing the Augmented Surface Method that there is good theoretical ground for concluding that the true "coefficient of augmentation" is less than the theoretical, and may well be unity. All things considered, Froude's position as to the nature of skin resistance appears, so far as our present knowledge goes, entirely justifiable.

The skin resistance of a ship will then be expressed by the same formula as the skin resistance of a plane area. That is, if R_s denote skin resistance in pounds, S the wetted surface in square feet, and V the speed in knots, $R_s = fSV^n$, where f and n are semi-constants, changing somewhat with the nature and dimensions of the wetted surface.

Constants.

Since we must depend upon difficult and costly experiments for the values of f and n, it is natural to find but few reliable determinations of them.

Table III. gives the constants used by Mr. R. E. Froude, the son and successor of Mr. William Froude. It is calculated from data published in a paper read before the Institution of Naval Architects in 1888.

Tables IV. and V. are calculated from tables of Tideman's coefficients, published in Busley's Schiffs-maschine. For ordinary use I consider Tideman's coefficients preferable, for reasons given below in discussing eddy resistance.

Eddy Resistance.

The eddy resistance is difficult to determine separately, since in the results of model experiments it appears combined with wave resistance. Froude concluded (about fifteen years ago) that for ordinary ships of good form the eddy resistance was usually about 8 per cent of the skin resistance.

For well-formed ships of the present day it probably seldom exceeds this value, and usually falls below it, so that 5 or 6 per cent would be a fair average allowance.

Now for full-sized ships the values of f, the coefficient of friction given by Tideman, are about 5 per cent greater than those used by Froude, so that if we are calculating resistance, it is preferable to use Tideman's Table, and not make a separate calculation for eddy resistance.

It is interesting in this connection to compare the resistances of eddy-making area and frictional surface. *Comparison of Eddy Making and Frictional Surface.*

Jœssel's formula for a plane of area A moving perpendicular to itself in sea-water becomes, for speed in knots,

$$R_e = 4.62 \, A V^2.$$

From this the eddy resistance of one square foot at 20 knots would be 1848 pounds. This is equivalent to the skin resistance at the same speed (for a ship 300 feet long) of about 770 square feet of wetted surface. While the eddy-making appendages of ships do not usually, and should never approximate to planes moving perpendicular to themselves, the necessity for so shaping them as to reduce eddy resistance to a minimum is obvious. This is a matter largely under control of the designer, and there is seldom any good excuse for exaggerated eddy resistance.

In this connection it should be said that if the lines of a ship aft are too full for her speed, large, unstable eddies are liable to appear, shifting from one counter to the other, and appearing and disappearing in a capricious and irregular way. *Excessive Eddy Making affects Steering.*

These suddenly formed and shifting eddies cause sudden and unsymmetrical changes of resistance, and render such a ship very difficult to steer.

A case in point is that of the short and full-sterned armoured ship *Ajax* of Her Majesty's Navy, which steered very wild at full speed until her run was lengthened and made finer.

§ 10. *The Law of Comparison.*

Why justly called Froude's Law.

Before completing the description of Froude's Method, I shall deduce and discuss what is commonly called Froude's Law, or the Law of Comparison. The formula of mechanics which is the most general expression of Froude's Law was discovered long before Froude. Its successful application to the comparison of the resistances of ships and their models is, however, due entirely to the late Mr. William Froude, and indeed Froude appears to have attained this result without any knowledge of the general formula alluded to above.

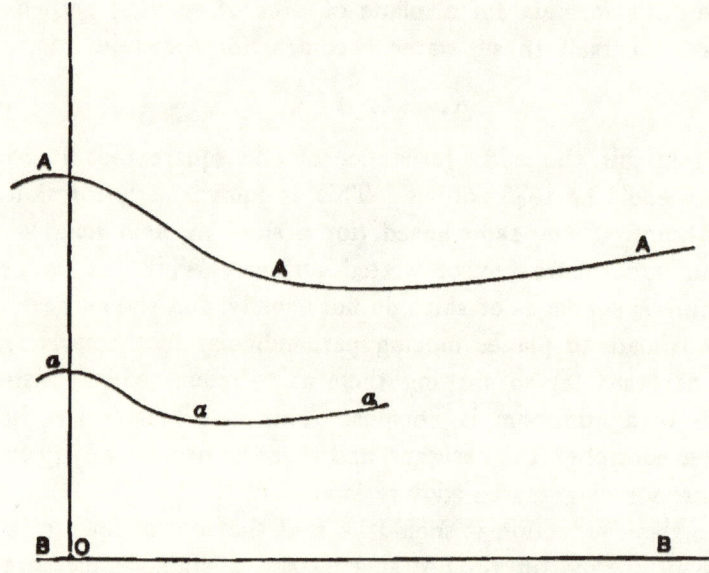

Fig. 8. — FOR LAW OF COMPARISON.

Deduction of Principle of Law of Comparison.

Referring to Figure 8, let BB be a fixed horizontal plane, and aaa, a stream line in a steady stream of perfect fluid. Suppose the upper surface of the fluid exposed to a constant pressure — as, for instance, that of the atmosphere — which denote by π. Let p denote pressure due to fluid alone; that is, the actual pressure, less the constant π.

§ 10. THE LAW OF COMPARISON.

Let v denote velocity of flow in feet per second.
Let z denote elevation above BB.
Let w denote weight per unit volume of fluid.
Then, at any point along the stream line, we have, from the steady motion formula,

$$\frac{p}{w}+\frac{v^2}{2g}+z=\text{a constant}=\frac{p_0}{w}+\frac{v_0^2}{2g}+z_0, \qquad (1)$$

where $p_0, v_0,$ and z_0 refer to the point of the stream line immediately above the origin O.

Suppose, now, the fluid removed and replaced by a second, whose pressure, velocity, etc., are denoted by large letters. Then for a stream line AAA in the second fluid,

$$\frac{P}{W}+\frac{V^2}{2g}+Z=\frac{P_0}{W}+\frac{V_0^2}{2g}+Z_0. \qquad (2)$$

Reducing (1) and (2),

$$\frac{p-p_0}{w}+\frac{v^2-v_0^2}{2g}+z-z_0=0, \qquad (3)$$

$$\frac{P-P_0}{W}+\frac{V^2-V_0^2}{2g}+Z-Z_0=0. \qquad (4)$$

Now let AAA be simply an enlargement of aaa, say n times as great.

Since the motions are similar,

$$v:v_0=V:V_0.$$

Whence,

$$1-\frac{v_0^2}{v^2}=1-\frac{V_0^2}{V^2}. \qquad (5)$$

Multiplying (3) by n and rewriting,

$$n\frac{p-p_0}{w}+\frac{nv^2}{2g}\left(1-\frac{v_0^2}{v^2}\right)+n(z-z_0)=0, \qquad (6)$$

and from (4),

$$\frac{P-P_0}{W}+\frac{V^2}{2g}\left(1-\frac{V_0^2}{V^2}\right)+Z-Z_0=0. \qquad (7)$$

Now, whatever the value of n,

$$Z = nz; \quad Z_0 = nz_0.$$

We are still free to give n a value in terms of v and V.

Let
$$n = \frac{V_0^2}{v_0^2} = \frac{V^2}{v^2}.$$

Subtracting (6) from (7), and cancelling terms which have become equal on the above supposition, we have

$$\frac{P - P_0}{W} = n\frac{p - p_0}{w}. \tag{8}$$

If, now, the stream lines are around some solid, they will at a great distance from the solid be parallel straight lines. The removal of the origin to such a distant point in the case of the first fluid, and to a point n times as far in the case of the second, does not change any of the previous results.

Suppose this done; then $\frac{P_0}{W}$ and $\frac{p_0}{w}$ will simply denote depths below the surface. Since the second fluid is on n times the scale of the first, undisturbed depths in the second will be n times undisturbed depths in the first. Therefore

$$\frac{P_0}{W} = n\frac{p_0}{w}. \tag{9}$$

Combining (8) and (9),

$$\frac{P}{W} = n\frac{p}{w}. \tag{10}$$

Application to Ship and Model. Apply now this result to the case of a ship and its model, supposed floating in streams of fluids of different densities.

Let L, B, H, the length, breadth, and draught of the ship, be n times l, b, h, the corresponding dimensions of the model. Let D' and d' denote the volumes displaced by ship and model respectively. Then $D' = n^3 d'$, since the volumes of

§ 10. THE LAW OF COMPARISON. 31

similar solids are as the cubes of their linear dimensions. Let V and v denote the normal velocities of the streams past ship and model, connected by the relation $V = v\sqrt{n}$.

Now if the model and ship, which are similar in every respect, produce similar stream lines, Equation 10 above $\left(\dfrac{P}{W} = n\dfrac{p}{w}\right)$ applies to the pressures at corresponding points of the similar stream lines.

If ds denote an element of surface of the model, and dS the corresponding element of the ship, evidently $dS = n^2 ds$. Let θ denote the inclination of the normals at ds and dS to the fore and aft vertical plane.

Let R_r, r_r denote the part of the ship and model resistances respectively due to the stream-line pressures. Evidently

$$r_r = \int p \cos\theta\, ds, \quad R_r = \int P \cos\theta\, dS,$$

the integration extending over the whole immersed surface in such case.

Now since $P = n\dfrac{W}{w}\cdot p$, and $dS = n^2 ds$,

$$R_r = n\dfrac{W}{w}\int p \cos\theta\, n^2 ds = n^3 \dfrac{W}{w}\int p \cos\theta\, ds = n^3 \dfrac{W r_r}{w};$$

or, $$\dfrac{R_r}{r_r} = \dfrac{W}{w n^3} = \dfrac{W}{w}\cdot\dfrac{D'}{d'}.$$

Now, $WD' =$ weight displaced by ship,

$wd' =$ weight displaced by model;

whence, $$\dfrac{R_r}{r_r} = \dfrac{\text{displacement of ship}}{\text{displacement of model}} = \dfrac{D}{d}.$$

These ship and model resistances are to each other in the ratio of the displacements, not at the same speed, but at speeds connected by the relation $V^2 = nv^2$, where n is the ratio between linear dimensions of ship and model. These

Corresponding Speeds.

speeds V and v are called the Corresponding Speeds of ship and model.

Now, $\quad V^2 = nv^2, \; n = \dfrac{V^2}{v^2},$

$$L = nl, \quad n = \dfrac{L}{l};$$

whence, $\quad \dfrac{V^2}{v^2} = \dfrac{L}{l}; \text{ or, } \dfrac{V^2}{L} = \dfrac{v^2}{l} = c^2, \text{ say}.$

Speed Length Ratio. At corresponding speeds, then, the value of c is the same for both ship and model. The quantity denoted by c is called the Speed Length Constant or Speed Length Ratio, the latter name being preferable since c is not a constant.

Assumptions of Froude's Law. In the deduction of the law of comparison two principal assumptions were made.

1. That we were dealing with a perfect fluid.
2. That the stream lines past models and their corresponding ships were similar — differing only in scale.

Justification of Froude's Law. We know that water is so nearly a perfect fluid as to justify the first assumption. The second assumption requires some evidence to justify it. The elder Froude made numerous experiments upon models, similar, but differing in size, and found that so far as careful observation could establish, the wave surfaces, and hence the stream lines, were similar at corresponding speeds. He found also that the Law of Comparison applied in such cases.

Froude recognised, however, that experiments with models, even though one were double the size of another, could hardly be considered conclusive in extending the law to full-sized ships.

Accordingly with the assistance of the Admiralty he carried out towing experiments to determine an actual curve of resistance for the *Greyhound* — a ship 172 feet long and of more than 1000 tons' displacement.

Figure 9 shows a comparison between the curve of resistance obtained by towing the ship and that obtained by calculating the skin resistance and deducing the residuary resistance by the Law of Comparison from that of a model 10¾ feet long. Figure 9 speaks for itself.

So far as I am aware the *Greyhound* experiments have not been repeated with any other ship, but the wave patterns of many men-of-war have been compared with those of their models at corresponding speeds by the Froudes, and appear to have been similar, so far as could be established by close observation.

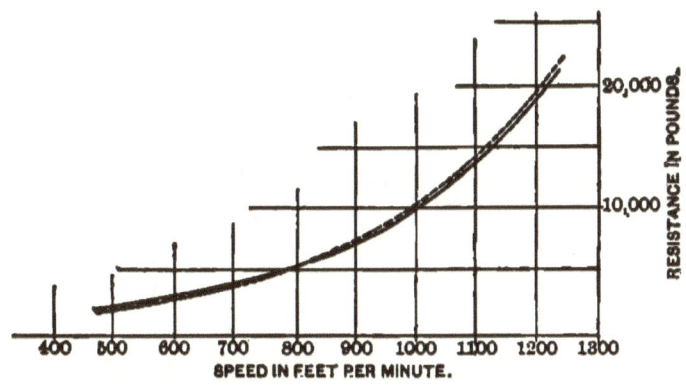

Full line — Actual resistance.
Dotted Line — Resistance as deduced from that of model.

Fig. 9. — ACTUAL RESISTANCE OF GREYHOUND AND RESISTANCE AS DEDUCED FROM THAT OF MODEL.

All things considered, we appear fully justified in accepting the Law of Comparison.

It will be well at this point to try and learn something of the law followed by a resistance which satisfies the Law of Comparison. Laws of Resistance which satisfy Froude's Law.

Suppose we have a resistance of such a simple nature that for similar ships of different sizes, at varying speeds, it is always expressed by the formula $b \cdot D^x V^y$, where b, x, and y are constants. Let R' denote this resistance for a full-sized ship, and r' the same resistance for a model.

Then $\qquad R' = bD^x V^y,$

$\qquad\qquad r' = bd^x v^y.$

Now if V and v are corresponding speeds,

$$\frac{V}{v} = \left(\frac{L}{l}\right)^{\frac{1}{2}} = \left(\frac{D}{d}\right)^{\frac{1}{6}}.$$

since $\qquad \dfrac{L}{l} = \left(\dfrac{D}{d}\right)^{\frac{1}{3}},$

also at corresponding speeds $\dfrac{R'}{r'} = \dfrac{D}{d}.$

Hence, $\qquad \dfrac{R'}{r'} = \left(\dfrac{D}{d}\right)^x \left(\dfrac{V}{v}\right)^y = \dfrac{D}{d};$

or, $\qquad \left(\dfrac{D}{d}\right)^{x-1} \times \left(\dfrac{V}{v}\right)^y = 1 = \left(\dfrac{D}{d}\right)^{x-1} \left\{\left(\dfrac{D}{d}\right)^{\frac{1}{6}}\right\}^y;$

whence, $\qquad \left(\dfrac{D}{d}\right)^{x + \frac{y}{6} - 1} = 1.$

Now $\dfrac{D}{d}$ is not equal to 1, so $\left(\dfrac{D}{d}\right)^{x + \frac{y}{6} - 1}$ can equal 1 only if

$$x + \frac{y}{6} - 1 = 0.$$

Hence if a resistance, expressed by the formula $bD^x V^y$, satisfies the Law of Comparison, we must have $x + \dfrac{y}{6} = 1$.

It is interesting to note the possible cases for integral values of y.

They are shown below.

Value of y.	Value of x.	Law of Resistance.	Value of y.	Value of x.	Law of Resistance.
0	1	bD	4	$\frac{1}{3}$	$bD^{\frac{1}{3}} V^4$
1	$\frac{5}{6}$	$bD^{\frac{5}{6}} V$	5	$\frac{1}{6}$	$bD^{\frac{1}{6}} V^5$
2	$\frac{2}{3}$	$bD^{\frac{2}{3}} V^2$	6	0	$b V^6$
3	$\frac{1}{2}$	$bD^{\frac{1}{2}} V^3$			

§ 10. THE LAW OF COMPARISON. 35

y cannot be $=0$, for we should then have a resistance which did not increase with the speed; nor can it be negative, for that would involve a resistance decreasing with the speed.

Also y cannot be 6, involving a resistance which does not increase with size; nor can it be greater than 6, for that would involve a resistance diminishing with the size for unchanged speed.

We know that eddy resistance varies as the square of the speed; and since it follows the Law of Comparison, it must follow the law,

$$R_e = aD^2 V^2.$$

This result could also be demonstrated independently of the Law of Comparison.

We shall see later that the wave resistance does not follow so simple a law as to be exactly capable of expression by such a formula as

$$R_w = aD^x V^y.$$

An approximate expression in this form is possible, however, which is applicable to many cases.

Since wetted surface varies as the square of the linear dimensions, or as $D^{\frac{2}{3}}$, the skin friction for similar ships of any size follows the law,

$$R_s = aD^{\frac{2}{3}} V^{1.83}.$$

This does not satisfy the Law of Comparison, and this is the reason why in Froude's Method the skin resistance must be calculated.

The fact that the skin resistance does not satisfy Froude's Law is of importance in the extension of Froude's Method to the powering of ships. Its bearing will be considered in the proper place.

§ 11. *Froude's Method* (*completed*).

Apparatus Necessary.
I am now in a position to complete the description of Froude's Method. For a full description of the apparatus and methods used by Froude, the reader should consult his papers read before the Institution of Naval Architects.

Without going into technical details it may be said that the models used are usually made of paraffin, being cast to approximate shape, and finished to exactly reproduce the ship's form. They are from 9 to 25 feet long. A good and very common working length is 12 feet. When of paraffin the shell is an inch or so thick. The models are ballasted by weights or shot to any desired water line.

The tank in which the models are towed should be 300 or 400 feet long, about 30 wide, and at least 10 or 12 deep.

The towing and recording apparatus used in resistance work is fitted to tow the model steadily at any desired speed, and to record the speed and the corresponding resistance. There are many practical difficulties in the way of obtaining accurate results, and to secure reliable data there is needed refined apparatus and great care and skill in handling it.

Example of Froude's Method.
Supposing all difficulties overcome, we can plot the results of towing experiments upon a model in the shape of a curve such as AAA in Figure 10, showing the resistances of the model plotted upon speeds as abscissæ. This curve represents the total resistance, made up, as we know, of skin resistance, eddy resistance, and wave resistance.

We also know that the two latter alone — constituting the residuary resistance — follow the Law of Comparison.

The first step, then, is to deduct from the total resistance the skin friction, which is calculated (as, for instance, by the use of Table IV.).

Setting down the skin friction from the curve AAA in Figure 10, we obtain the curve BBB, representing the residuary resistance of the model.

FROUDE'S METHOD.

Now we know from the Law of Comparison that this curve also represents the residuary resistance of the ship, provided the scales of speed and resistance are suitably changed.

In the case shown by Figure 10 the model was $\frac{1}{16}$ the size of the ship; hence corresponding speeds of ship and model are in the ratio $\sqrt{16} : 1 = 4 : 1$ and residuary resistances at corresponding speeds are in the ratio $16^3 : 1 = 4096 : 1$.

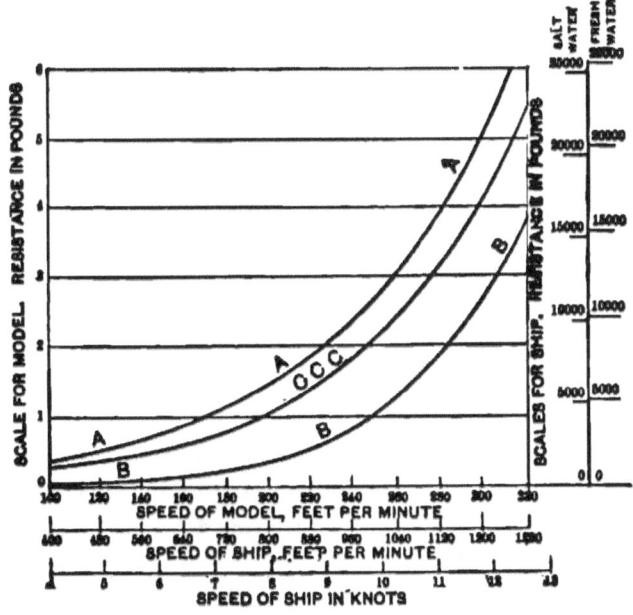

Fig. 10.

Drawing in the scales for the ship as shown, the curve BBB represents the residuary resistance of the ship in either fresh or salt water, according to the scale used.

(Salt-water resistance) = 1.026 (fresh-water resistance).

It is now necessary to calculate the skin resistance of the ship, and set it up above BBB to obtain the curve CCC, which represents the total resistance of the ship.

The curves of Figure 10 are those for H. M. S. *Greyhound*, given by Froude. Figure 9 shows the close agree-

Special Experiments for Skin Resistance in Accurate Work.

ment between the curve *CCC*, determined as just described, and the actual curve, determined by towing experiments upon the ship. While I have for clearness of description assumed Tables IV. and V. applicable to the case, the skin-resistance constants, used by Froude in this case, were obtained by special experiments.

In accurate tank work the skin resistance of a model should always be deduced from direct experiments upon plane surfaces of the same length and area and nature as the model's surface. This is because an error, small in the case of the model, becomes much amplified in proceeding to the full-sized ship.

Value of Froude's Method. Froude's Method is at present by far the most accurate and reliable known for determining the resistance of a ship. It should be pointed out, however, that it is not, does not pretend to be, and cannot be, exact.

The skin resistance is calculated, and the coefficients used in calculations must often be different from the actual coefficients. Our knowledge of the exact frictional value of painted surfaces is limited, and the standard but twenty-year old experiments of Froude need supplementing by exhaustive experiments confined to painted iron surfaces in various conditions, and extending, if possible, to a length of plane greater than 50 feet. Such experiments would, I believe, show that Froude's and Tideman's coefficients are thoroughly reliable for working approximations, but would enable a somewhat closer approximation to be made in many cases.

§ 12. *Phenomena of Waves produced by Ships.*

Need of Formula for Wave Resistance. It has been seen that the skin resistance can be calculated with an accuracy sufficient for practical purposes. The combined wave and eddy resistance can be very closely determined by Froude's Method. In the absence of tank facilities a satisfactory allowance can be made for eddy

§ 12. PHENOMENA OF WAVES PRODUCED BY SHIPS.

resistance, which should never be more than a very small percentage of the total. Since but few of us are favoured with facilities for model tank experiments, an approximate method for calculating the wave resistance in a given case is much needed.

Of the nature, laws, and phenomena of wave resistance we can learn something from pure theory, and something from results of experiments given by the Froudes in various papers before the Institution of Naval Architects.

Cause and Genesis of Waves. — We have seen that in fluid streaming past a submerged body there are changes from the normal pressure and velocity, the pressure in the stream being greater than the normal near the bow and stern, and less than the normal amidships. Tendency toward change of pressure must, of course, exist in a stream past a floating ship. But the pressure at the surface must remain constant — being the pressure of the atmosphere.

Changes in velocity of flow appear; but at the surface, instead of changes of pressure, changes in the level of the water make their appearance; or, in other words, waves are produced.

If, as imagined by Froude, we could surround a ship by a thin sheet of rigid ice on the surface of the stream, extending to a great distance in all directions, there would be no changes in the level of the surface. There would be changes in pressure and velocity of the stream caused by the ship, but they would be confined to the vicinity of the vessel. When waves are produced upon a free surface, however, the case is different. The waves follow their natural tendency to spread, and much or all of the energy required to produce them cannot be returned to the ship.

Two Separate Systems generated by a Ship. — Consider now a ship with a long parallel middle body. The stream-line pressures at the bow will cause waves. The disturbance of the surface will spread away from the ship, and if the parallel middle body is long enough there will be at the stern no remaining disturbance due to the bow. But as the stream lines close in aft, there will be new variations of pressure which must cause new changes of water level, producing a new wave system.

It is evident, then, that a ship tends to originate two separate wave systems, one forward and one aft

Fig. 11. — BOW WAVE SYSTEM.

I shall distinguish them as the "natural" bow wave or bow wave system, and the natural stern wave or stern wave system.

Bow Wave. Figure 11 shows the features of a bow wave system. For clearness, vertical heights have been exaggerated.

NOTE. — From a paper by Froude, read before the I. N. A.

We observe a diverging series of pronounced crests spreading away from the ship. The diverging crests are parallel to each other. The inclination of each crest to the line of advance of the ship is about double that of the "line of divergence," or the line separating the disturbed from the undisturbed water.

It is to be noted that the line of divergence is straight or nearly so.

§ 12. PHENOMENA OF WAVES PRODUCED BY SHIPS.   41

It is also seen by inspection that each of the diverging crests appears to form the end of a "transverse" wave whose crest is nearly perpendicular to the line of advance, and shows in the figure against the side.

The natural stern wave system, which tends to form at the stern, is similar in character to the bow wave. As a rule, however, we have at the stern more or less disturbance due to the bow wave, and this modifies the natural stern wave in a manner which I shall discuss later. It should be explained, *Stern Wave.*

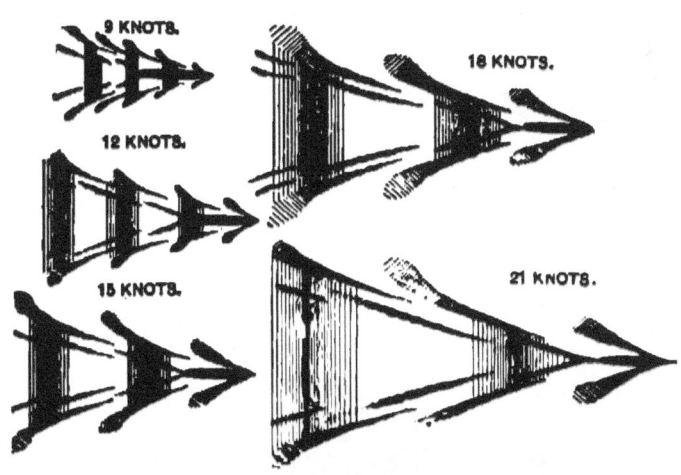

N.B. *Position of Wave Crests indicated by shading.*

Fig. 12a. — PLAN OF WAVE SYSTEM MADE BY 83 FEET LAUNCH AT VARIOUS SPEEDS.

too, that in ships with little or no parallel middle body the whole of the fore body comes into play in generating the bow wave, and the whole of the after body in generating the stern wave.

As a further illustration of the features of the waves produced by a ship, I may refer to Figures 12*a* and 12*b*. *Wave Systems in Practice.*

NOTE. — From a paper, read by the younger Froude before the I. N. A.

These show in plan various wave systems.

In this connection it should be remarked that, except at high speeds or with very bluff vessels, the wave systems are seldom so clearly defined in practice as shown by Figures 12a, etc.

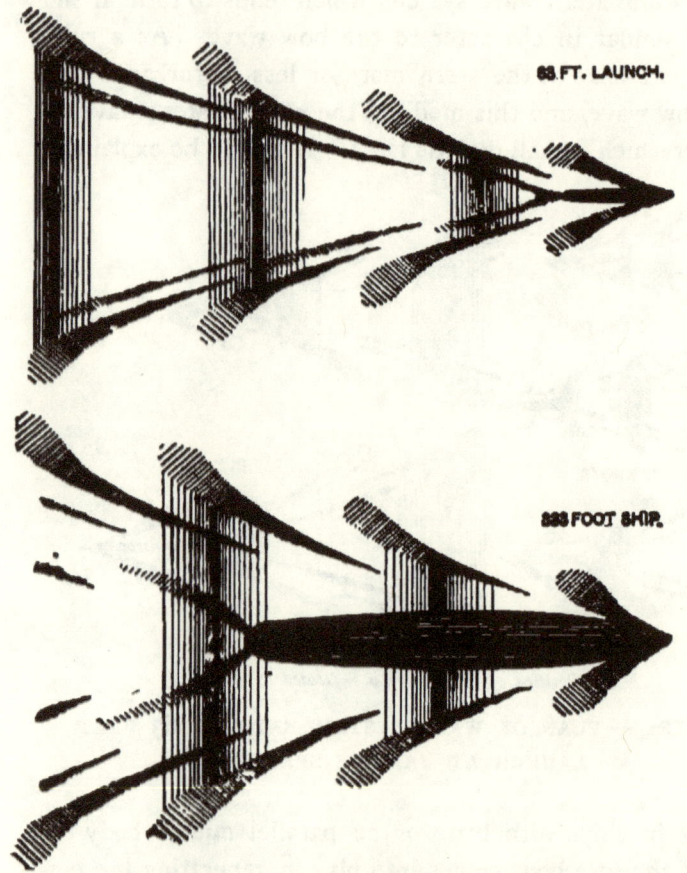

N.B. *Position of Wave Crests indicated by shading.*

Fig. 12b. — PLANS OF WAVE SYSTEMS MADE BY DIFFERENT VESSELS AT 18 KNOTS SPEED.

The transverse crests are not well marked, and the bow wave especially often appears at low speeds to consist entirely of the diverging crests. At high speeds, however, when the wave resistance becomes important, the features which have been described show plainly near the ship, and it is only at some distance astern that the transverse crests apparently disappear.

The "diverging wave system" and "transverse wave system" are sometimes spoken of as if they were separate and disconnected, but this appears mechanically and physically impossible. They are essentially one. The diverging crests appear always to form the end of the transverse waves, though in some cases the transverse wave is all end, so to speak, extending within the line of the diverging crests but a very small distance toward the ship.

The most beautiful large-scale wave systems of my acquaintance are those which spread aft from the paddle wheels of a powerful paddle steamer. Their transverse and diverging features are plainly marked, the former being proportionately more strongly accented than in the case of bow and stern wave systems.

§ 13. *Properties of Trochoidal Waves.*

In this section I shall state certain properties of trochoidal waves, which may be found demonstrated by Rankine or other writers upon the subject.

Rankine's trochoidal wave theory is founded upon the assumption that in deep-water waves the particles of water affected by the wave revolve uniformly in circular orbits. The theory is practically exact as applied to a system of uniform regular waves, indefinite in number, and advancing steadily in a direction perpendicular to their crest lines. *(Genesis of Trochoidal Waves.)*

Let w denote the weight per cubic foot of the water. Let l denote the length of the wave (*i.e.* the distance from one crest to the next), and H the height of the wave from crest to hollow, both l and H being measured in feet.

Consider a solid mass of water, extending to the bottom,

of breadth in the direction of the crests $=b$, and of length in the direction of propagation of the waves $=l$.

Then we have the following properties:

Kinetic Energy.
1. The kinetic energy of the mass, due to the rotation of its particles, remains constant while the wave passes, and is equal in foot pounds to $\frac{1}{16} wblH^2$.

Potential Energy.
2. The mean centre of gravity is raised above the position it would occupy if the water were at rest, so that the mass has a certain amount of potential energy. This potential energy remains the same during the passage of the wave, and is also equal in foot pounds to $\frac{1}{16} wblH^2$.

Transmission of Energy.
3. The potential energy remains constant because, while the mass of water considered is constantly receiving energy from the water behind, it delivers energy at the same rate to the water in front. During the passage of one wave (*i.e.* during one wave period) the amount of energy so transmitted is equal to the constant potential energy of the mass $=\frac{1}{16} wblH^2$.

Compound Waves.
4. Let H_1, H_2, be the heights of two trochoidal wave series of the same length $=2\pi R$, where R is the radius of the rolling or generating circle. Then if one series be superposed upon the other with an interval between crests $=a$, we shall have resulting a single series of trochoidal waves of a height H such that

$$H^2 = H_1^2 + H_2^2 + 2H_1H_2 \cos\frac{a}{R},$$

and having the same speed of advance as each of the components.

Speed of Advance.
5. If v denote the speed of advance in feet per second of a trochoidal wave, and R the radius of the rolling or generating circle, then l, the length of the wave, $=2\pi R$ and,

$$v^2 = gR = \frac{gl}{2\pi}.$$

§ 14. Deduction of Law of Wave Resistance.

It is found that the distance fore and aft between successive crests of the bow wave system is, as nearly as possible, the same as that between successive crests of a trochoidal system, with speed of advance equal to the speed of the ship. *Trochoidal Substitute for Bow Wave System.*

Also at high speeds, when the wave resistance becomes a large portion of the total, the transverse waves which appear strongly resemble short lengths of trochoidal waves.

We may then for convenience take the energy of the bow wave system as concentrated in the transverse waves, supposed to be trochoidal.

Advancing into still water, such a system carries its potential energy with it; but the kinetic energy of the new particles, constantly set in motion, must be derived from work done by the wave resistance of the ship.

Let b denote the breadth at a given point from the line of divergence on the starboard side to that on the port side. Let H denote the height at that point of the uniform trochoidal wave, which is supposed to replace the somewhat irregular actual wave. Let R_w denote the wave resistance at speed V. Then $R_w V$ denotes the work done against wave resistance in unit time, and must equal the kinetic energy generated in the water per unit time. Now the time required to traverse the distance l (= one wave length) is $\dfrac{l}{V}$. Then in the time required to traverse one wave length, *Resistance due to Bow Wave System.*

Work done by ship $= R_w V \times \dfrac{l}{V} = R_w l$;

Kinetic energy generated in water $= \tfrac{1}{16} w b l H^2$.

Whence, $R_w l = \tfrac{1}{16} w b l H^2$ or $R_w = \tfrac{1}{16} w b H^2$.

b changes but little with the speed; hence we may reasonably conclude that as the speed changes the wave resistance varies as H^2.

Now H is the height of the imaginary trochoidal wave, supposed to replace the actual wave, and cannot differ much from the mean height of the actual wave. The exact determination of H in a given case is impracticable, but, as will be seen, it is not necessary for the purpose in hand.

Stern Wave.

The preceding applies directly to the bow wave, and would apply to the natural stern wave. But the actual stern wave is the resultant of the natural stern wave and of part of the bow wave.

Although the positions of the first crests of the bow and stern wave systems of a given ship change with the speed, there will at a given speed be a fixed distance between them. Denote this distance, called the wave-making length, by s, and suppose $s = mL$, where m is a coefficient varying slightly as the speed changes.

Let R denote the radius of the rolling circle of the trochoidal bow wave. Then its length $= 2\pi R$; and since $v^2 = gR$, its length $= \dfrac{2\pi v^2}{g}$.

Now the distance from the natural position of the first stern wave crest to the bow wave crest next ahead of it is evidently the remainder after dividing s; the distance from first bow crest to first stern crest, by $\dfrac{2\pi v^2}{g}$, the length from crest to crest of the bow wave.

Referring to the compound wave formula, it is seen that this remainder corresponds to a in that formula.

Now let $s = (n+q)\dfrac{2\pi v^2}{g}$, where n is a whole number, and q a fraction.

Then a in the compound wave formula $= q \cdot \dfrac{2\pi v^2}{g}$; and since $R = \dfrac{v^2}{g}$, $\dfrac{a}{R} = 2\pi q$,

§ 14. DEDUCTION OF LAW OF WAVE RESISTANCE.

$$\cos \frac{a}{R} = \cos 2\pi q = \cos(2\pi q + 2\pi n)$$

$$= \cos \frac{2\pi(n+q)\frac{v^2}{g}}{\frac{v^2}{g}} = \cos \frac{2\pi R(n+q)}{\frac{v^2}{g}}$$

$$= \cos \frac{s}{\frac{v^2}{g}} = \cos \frac{mL}{\frac{v^2}{g}}$$

$$= \cos \frac{gm}{\frac{v^2}{L}}, \text{ where } v \text{ is in feet per second,}$$

$$= \cos \frac{gm}{\frac{V^2}{L}} \left(\frac{3600}{6080}\right)^2, \text{ where } V \text{ is in knots.}$$

Now $\frac{V^2}{L}$ is the speed length ratio squared denoted by c^2.

Substituting for g its value, and changing from circular or absolute measure to degrees, we have $\cos \frac{a}{R} = \cos \frac{m}{c^2} \times 646°$.

Now let H_1 denote the height of the bow wave at a given breadth b, H_2 the height of the natural stern wave at the same breadth, and kH_1, the height of the bow wave when it has passed aft to the point where the natural stern wave has the breadth b. Let H'_2 denote the height of the actual compound stern wave at the breadth b. Resistance due to Stern Wave.

Then from the preceding

$$H'^2_2 = k^2 H_1^2 + H_2^2 + 2kH_1H_2 \cos \frac{m}{c^2} 646°.$$

The resistance due to the actual stern wave will be proportional to H'^2_2, and the resistance due to the portion of the bow wave, not compounded with the stern wave, will be proportional to $H_1^2 - k^2 H_1^2$.

Total Wave Resistance.

Then will the total wave resistance be proportional to

$$H_1^2 - k^2 H_1^2 + H_2'^2$$

$$= H_1^2 - k^2 H_1^2 + k^2 H_1^2 + H_2^2 + 2 k H_1 H_2 \cos \frac{m}{c^2} 646°$$

$$= H_1^2 + H_2^2 + 2 k H_1 H_2 \cos \frac{m}{c^2} 646°.$$

This formula is not in shape for practical use, because of uncertainty as to the values of H_1, H_2, k, and m in a given case; but it expresses, I believe, the true law of wave resistance, and useful practical conclusions, confirmed by experiment, can be drawn as to the characteristics and mode of variation of the unknown quantities.

It was first given in a slightly different shape by the younger Froude.

§ 15. *Laws of Variation of Wave Resistance.*

Let us inquire first into the connection between H_1 and H_2 and the speed. We know that in perfect stream motion the excess pressure, at a point near the bow, for instance, will vary as the square of the speed.

Height of Waves as affected by Speed.

In the imperfect stream motion which exists, part of the excess pressure will be devoted as before to the production of stream line acceleration. The remainder will leak away, as it were, and be absorbed in raising the level of the water in the vicinity. Now if the resultant rise of level — or H_1 the height of the bow wave — were always proportional to the excess stream line pressure, we should have H_1 varying as V^2, and H_1^2 as V^4. The same reasoning applies to H_2^2.

It appears entirely probable that at moderate speeds when the water has time to obey the stream line pressures, so to speak, and the vertical motions are small, the heights of the natural bow and stern waves do vary somewhat closely as the square of the speed. As the speed increases, involving greater wave height and vertical motion of the water, it

§ 15. LAWS OF VARIATION OF WAVE RESISTANCE.

appears that the wave height should not increase so fast as the square of the speed. Now H_1^2 and H_2^2 are proportional to the natural bow and stern wave resistances respectively.

If the above reasoning is sound, it follows that the natural bow and stern wave resistance will vary at low and moderate speeds, as the fourth power of the speed, but that as the speed increases, the index will steadily fall off.

Consider next the coefficient k in the expression for wave resistance. *Variation of k.*

It is a matter of common experience that at low speeds, when the length of the bow wave is small in comparison with the length of the ship, the bow wave has practically subsided close to the ship by the time the stern is reached, and hence can produce little or no effect upon the natural stern wave.

As, however, with increasing speed, the length of the bow wave increases, so that fewer crests appear in the length of the ship, more and more of its energy is found in the vicinity of the stern — available as a component of the actual stern wave.

The general nature of variation of k is then obvious. k must be equal to zero at low speeds, and increases with the speed. The theoretical limit of k is unity, but it does not appear likely that for actual ships of the customary form k can ever exceed the value 0.5.

There is little to be said about m. It appears that the distance between the first bow wave crest and the first stern wave crest is, in all cases, somewhat greater than the length of the ship. Also, as is natural, this distance appears to increase slightly with the speed. For ships of ordinary form and speeds a fairly safe value for m appears to be about 1.10. It is to be regretted that the limited amount of data available prevents the deduction of a close approximation to m, but it is not a matter of serious practical importance. *Value of m.*

Having considered the component factors, we can now form some idea of the nature of a curve of wave resistance.

The wave resistance was taken as proportional to

$$H_1^2 + H_2^2 + 2kH_1H_2 \cos \frac{m}{c^2} 646°.$$

Wave Resistance Formula. Denote the resistance due to the natural bow wave by A^2V^4, and to the natural stern wave by B^2V^4. Then from what has been said, if R_w denote the wave resistance,

$$R_w = V^4(A^2 + B^2 + 2kAB \cos \frac{m}{c^2} 646°).$$

Features of Wave Resistance Curves. What would be the nature of a curve of wave resistance calculated from this formula and plotted on speeds as abscissæ?

At low speeds $k = 0$, nearly, and A and B are nearly constant. At such speeds, then, the wave resistance would vary as the fourth power of the speed, and would be expressed by $V^4 \times$ a constant.

As the speed increases, A and B fall off, while k ceases to be negligible and steadily increases. Now the term

$$2kAB \cos \frac{m}{c^2} 646°$$

is sometimes negative and sometimes positive.

If, then, the mean wave resistance equals $V^4(A^2 + B^2)$, the actual curve following the law,

$$R_w = V^4(A^2 + B^2 + 2kAB \cos \frac{m}{c^2} 646°),$$

will sometimes rise above and sometimes fall below the mean curve. This will give rise to "humps" and "hollows" so called in the curve of wave resistance.

As the maximum and minimum values of $\cos \frac{m}{c^2} 646°$ corresponding to $n \times 360°$ and $n \times (360° + 180°)$ (where n is an integer) are not spaced at equal intervals of speed, the intervals between successive "humps" will be greater and greater as the speed increases.

§ 15 LAWS OF VARIATION OF WAVE RESISTANCE. 51

As an illustration I refer to Figure 13, which shows a curve of values of $\cos \dfrac{m}{c^2} 646°$ plotted on speed. It is assumed that the ship is 300 feet long and that in this case $m = 1.05$. It should be remembered that $c = \dfrac{V}{\sqrt{L}}$, $c^2 = \dfrac{V^2}{L}$.

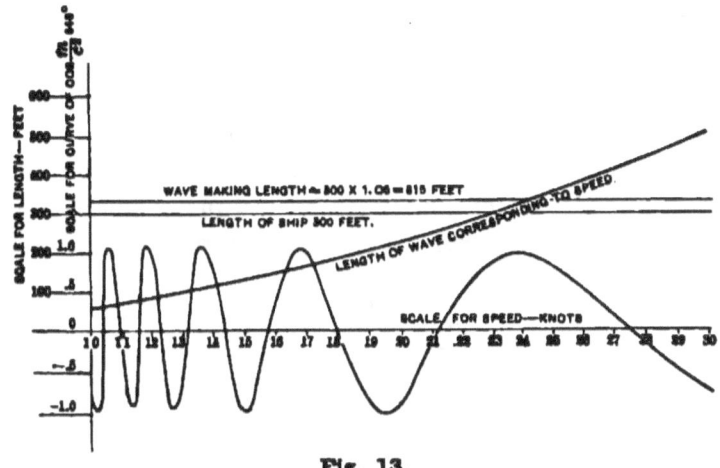

Fig. 13.

Figure 14 shows two curves of values of $\cos \dfrac{m}{c^2} 646°$ plotted on values of c as abscissæ.

For the full curve $m = 1.00$, and for the dotted curve $m = 1.10$.

These are values of m likely to be met with in practice.

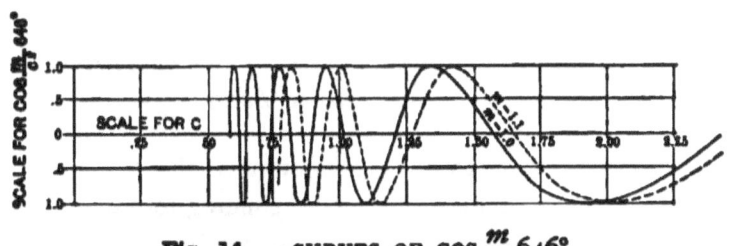

Fig. 14. — CURVES OF $\cos \dfrac{m}{c^2} 646°$.

The rapid fluctuations of the curves as c diminishes do not indicate features which would appear on a curve of wave resistance. For low speeds — say for c less than .6 — the coefficient k is practically zero, so it makes no difference what may be the value of cos $\frac{m}{c^2}$ 646°.

The first maximum value of cos $\frac{m}{c^2}$ 646° at which the value of k cannot be ignored is that found about $c = 1.00$.

Since at about this speed k is increasing, and A and B have usually not fallen off much, the "hump" found in this vicinity is a notable one. Few vessels, except torpedo boats, travel at a speed sufficient to surmount this hump.

Figure 13 also shows a curve showing for a ship 300 feet long the length of the wave produced plotted on values of v.

We have seen that, for speed in feet per second, if l denote the length of a wave of speed v,

$$v^2 = \frac{gl}{2\pi},$$

$$l = \frac{2\pi}{g} v^2 = \frac{2\pi}{g} \times \left(\frac{6080}{3600}\right)^2 v^2 = .5573\, v^2,$$

for speed in knots, and $g = 32.16$.

Whence

$$l = \frac{2\pi}{g} \times \left(\frac{6080}{3600}\right)^2 \times c^2 L = .5573\, c^2 L$$

$$= 167.19\, c^2 \text{ for } L = 300.$$

§ 16. *Results of Experiments on Wave Resistance and Approximate Laws.*

Having set forth in a general way the theory of wave resistance and deduced the corresponding formulæ, I propose now to quote certain experimental results bearing upon the matter.

Froude's Curves illustrating Interference.

Figures 15–17 show various curves of residuary resistance obtained by Froude from model experiments. The residuary

§ 16. RESULTS OF EXPERIMENTS. 53

resistance includes both eddy and wave resistance, but in the cases in question the eddy resistance must be so small that the curves may be said to represent practically the wave resistance alone.

Consider first Figure 15. This shows curves of residuary resistance for various speeds, plotted on length of parallel middle body. The lengths of entrance and of run and their shapes are identical, being always 80 feet, while the length of parallel middle body varies from zero to 340 feet.

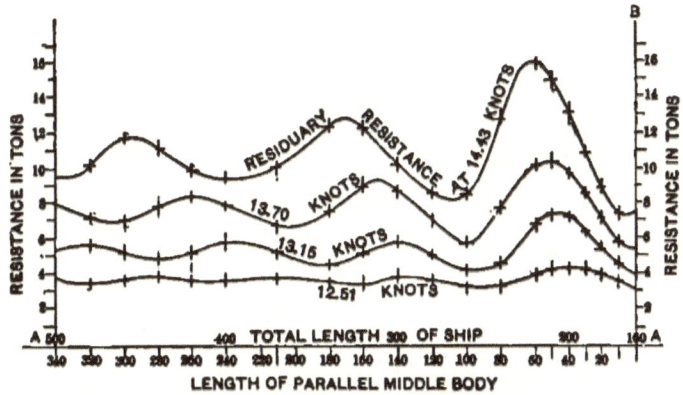

Fig. 15.—EFFECT OF ADDING PARALLEL MIDDLE BODY UPON RESIDUARY RESISTANCE.

These curves are worthy of close study.

The fluctuation of wave resistance due to the term in the formula $2\,kAB \cos \frac{m}{c^2} 646°$ is marked. It is to be noted that as the speed increases for the same length, the fluctuation increases — due to the increasing value of k. Also as the length is increased for a given speed, the fluctuation decreases, due to the fact that k falls off under these circumstances. This is of course because when the ship is lengthened the stern wave system is initiated at a greater distance aft of the bow wave and there is less interference between the two. If the length of parallel middle body were sufficiently great, the

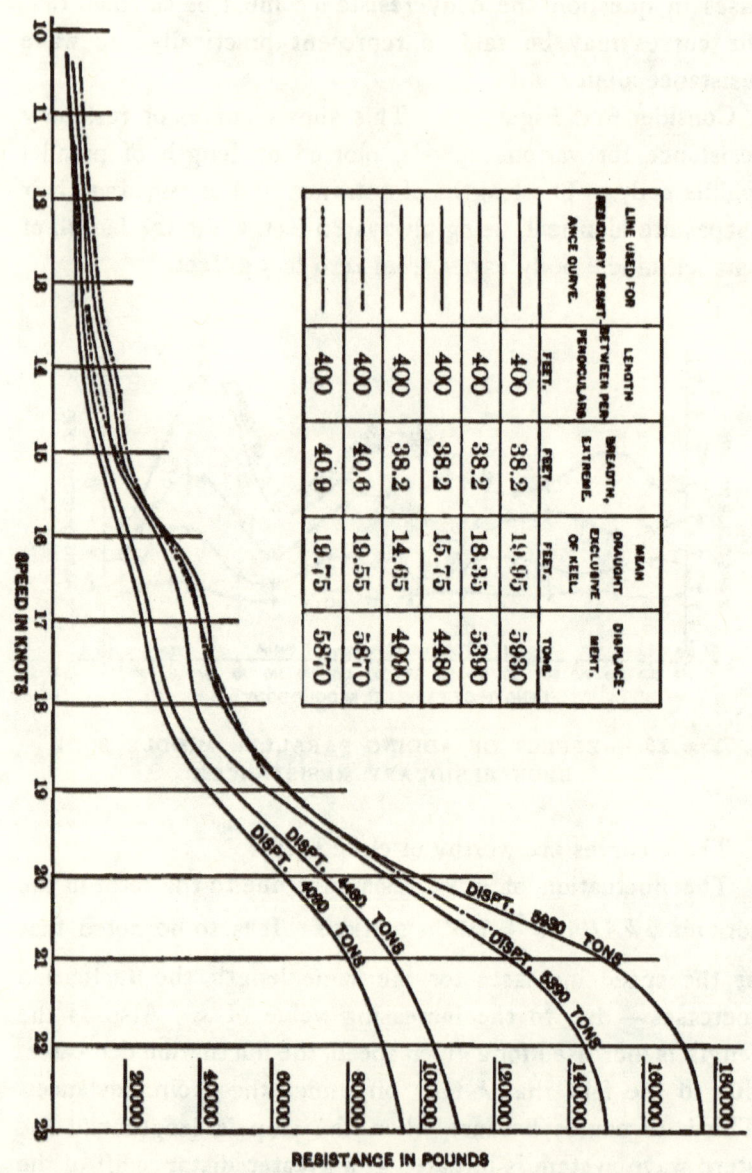

Fig. 16.—CURVES OF RESIDUARY RESISTANCE.

§ 16.   RESULTS OF EXPERIMENTS.   55

curve of wave resistance at a given speed, plotted on length of middle body, would finally cease to fluctuate, and become a level line.

Consider now Figure 16. It shows curves of residuary resistance of ships of the dimensions stated up to a speed of 23 knots. These curves show plainly the "humps" and

Curves of Residuary Resistance.

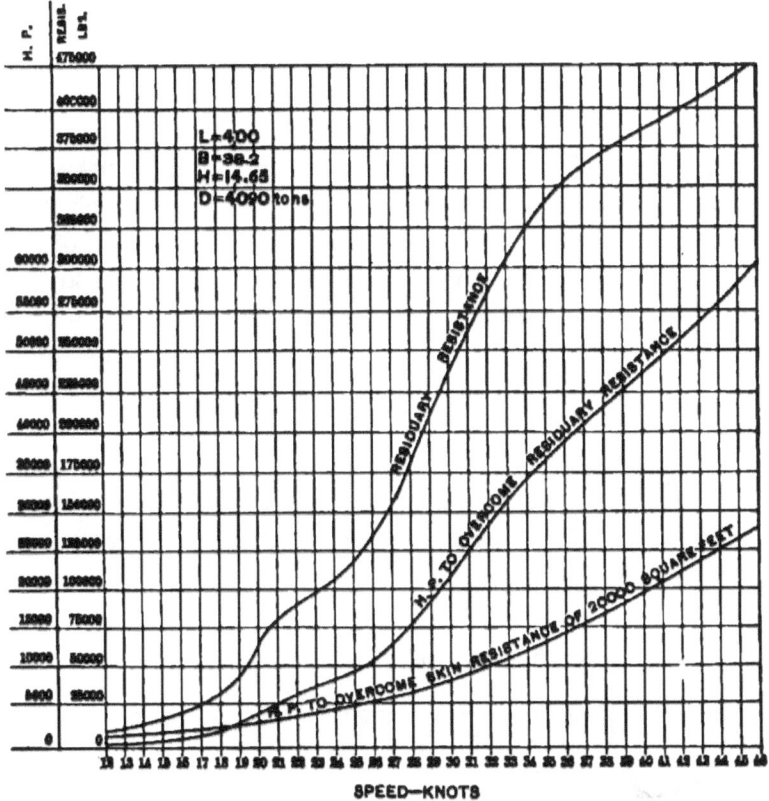

Fig. 17. — CURVES OF RESIDUARY RESISTANCE, ETC.

"hollows" which we have seen should characterise a curve of wave resistance.

These features are more clearly marked in the curves of Figure 16, which are deduced from model experiments, than

Residuary Resistance; Extreme Speed.

they would be in the case of actual ships properly designed to be driven at the speeds to which the curves extend.

Figure 17 shows one of the curves of Figure 16 extended to a speed of 46 knots. The curve of horse-power corresponding to the residuary resistance is also shown. The wetted surface of a ship of the dimensions and displacement of that in Figure 17 would be about 20,000 square feet.

The horse-power absorbed by the skin resistance of this amount of wetted surface is shown in Figure 17.

The paramount influence of skin resistance at low speeds and of wave resistance at high speeds is obvious.

While no ship of the size specified in Figure 17 has ever moved at anything like the extreme speeds to which the curves extend, it should be borne in mind that a speed of 40 knots for the 400-foot ship corresponds to that of 20 knots for the 100-foot torpedo boat, though the latter can attain the speed of 20 knots only by the adoption of a shape and proportions very different from that of the model of the 400-foot vessel.

A feature of Figure 17 worthy of note is that the "humps" and "hollows" of the curve of resistance are appreciably toned down in the curve of the horse-power necessary to overcome that resistance.

§ 17. *Simplification of Wave Resistance Formula.*

Complete Formula.

Let us return now to the wave resistance formula

$$R_w = V^4(A^2 + B^2 + 2kAB\frac{\cos m}{c^2} 646°).$$

This involves too many coefficients of uncertain and variable values for everyday use.

Omission of Last Term.

Obviously the first step in the direction of simplification will be to drop the last term, for the reason that k is quite small for the speeds attained in practice by ships of any size. This leaves us $R_w = V^4(A^2 + B^2)$.

§ 17.   WAVE RESISTANCE FORMULA.   57

Now we have seen that A and B are not constants, but must diminish with the speed.

Approximate Law.

So we cannot be sure in advance that this diminution is not so rapid that the best approximate formula for the wave resistance will be

$$R_w = V^3 \times \text{a constant}$$

instead of $R_w = V^4 \times \text{a constant}.$

I am of the opinion that the best working approximation is given by the formula

$$R_w = bV^4, \text{ where } b \text{ is a constant.}$$

To test the justness of this conclusion we may assume that the residuary resistance of Figure 17 is expressed by

Conclusions from Experimental Curve.

$$R_w = b_1 V^5,$$

or $R_w = bV^4,$

or $R_w = b_2 V^3.$

Then knowing the residuary resistance at each point from the curve, it is easy to plot curves of b_1, b, and b_2.

Figure 18 shows these curves. It is noted that all three curves show a marked lump at about 20 knots. It should be remembered, however, that the form and proportions of the ship were not suited to a speed so high as 20 knots, and that with a form suited to that speed, the excrescence would have been much less marked.

Below 12 knots the wave resistance is of little relative importance, and we need not concern ourselves at present about what goes on at speeds above 30 knots.

It is evident from Figure 18, speaking broadly (and bearing in mind that the lump about 20 knots is abnormal), that from 12 to 30 knots the curve of b_1 falls off steadily, and that of b_2 increases steadily, while that of b remains fairly constant.

Evidently the formula $R_w = bV^4$ is best adapted to this case, and so far as my experience goes the wave resistance is

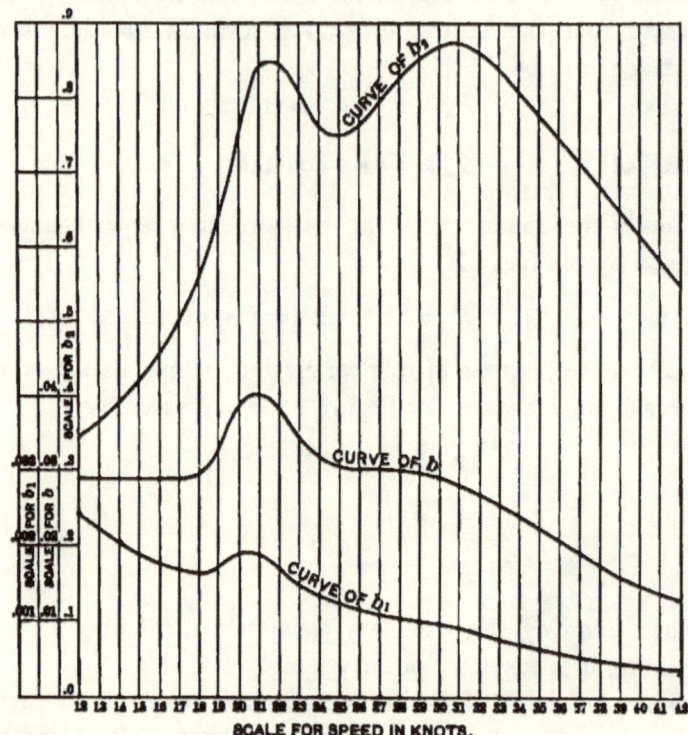

Fig. 18. — CURVES FOR LAW OF RESIDUARY RESISTANCE.

expressed with sufficient approximation by this formula up to speeds such that the speeds length ratio does not exceed 1.2.

§ 18. *Increase of Resistance in Shallow Water.*

Effect of Shallow Water upon Waves produced by Ship.   It is well known that the form of water waves changes in shoal water. According to the trochoidal theory the orbits of the particles cease to be circular in shoal water and become elliptical, the excentricity of the ellipse increasing with the shoalness of the water. Other changes in period, etc., take place as shown in Table VI., taken from a lecture on water waves by the late Richard Gatewood, naval constructor, U.S.N.

The character of the waves produced by a ship changes in a marked manner in shallow water, and their height is augmented. The result is increased resistance. This is an important point in considering the location of a measured course to try the speed of ships. It is reasonable to suppose that at depths where a trochoidal wave of length such as to travel at a given speed is practically unchanged from the deep water wave at the same speed the resistance of a ship at that speed will not be affected. Using Table VI., we have the following results:

Speed of Ship Knots.	Minimum depth for No Change in Resistance. FEET.
10	28
12	40
14	55
16	71
18	90
20	111
22	135
24	160
26	188
28	218
30	250

Depth of Water Appropriate to Various Speeds.

The increase of resistance due to shoal water appears liable to exaggeration.

Mr. R. E. Froude has stated (Transactions I. N. A., 1892) that for a 5000-ton ship at speeds no greater than 17 knots in water of 7 fathoms' depth the increase of resistance above that in deep water is but some 3 or 4 per cent.

§ 19. *Squat and Change of Trim.*

The line of flotation of a ship moving through the water is different from her line of flotation in still water.

We know that in perfect stream motion the pressures at *Squat.*

bow and stern are in excess, and the pressure amidships in defect. Since the sections amidships are fuller than those forward and aft, it would seem that the ship in motion should show a tendency to sink bodily in the water. This effect is produced to a slight extent, and is called "squat." Its result is to increase the wetted surface, and hence the skin resistance, but the increase is very small.

Change of Trim. The changes of pressure referred to above correspond to perfect stream motion, and being distributed nearly symmetrically forward and aft, cause little if any change of trim. But the stream line motion not being perfect and the excess pressure at the bow not being balanced by a similar pressure aft, a ship under way will nearly always change her trim more or less by the stern. The change is inappreciable in most cases, but with small high-speed boats is sometimes very marked.

Being an effect, and not a cause of resistance (except as it changes the shape, etc., of the under-water body), change of trim is not a matter of much importance in considering resistance.

§ 20. *Formula for Total Resistance.*

Before attacking the question of propulsion it will be well to summarise the results arrived at as to resistance.

Formula for Total Resistance. Combining the expressions for skin, eddy, and wave resistance, it appears that a comprehensive formula for total resistance would be

$$R = fSV^{1.83} + KV^2 + V^4\left(A^2 + B^2 + 2kAB \cos\frac{m}{c^2}646°\right).$$

Eddy Resistance. The second term on the right-hand side should be made small by careful design. Also it appears that by adopting values of f in Tideman's Table (which are nearly 5% larger than Froude's values), we make ample allowance for it. So exit eddy resistance, and we write

$$R = fSV^{1.83} + V^4\left(A^2 + B^2 + 2kAB \cos\frac{m}{c^2}646°\right).$$

§ 20. FORMULA FOR TOTAL RESISTANCE.

The second term now represents wave resistance.

We know that for low speeds — and by "low" I mean speeds for which the speed length ratio is not more than .5 to .6 — the wave resistance is seldom more than 10 % of the whole.

It is allowable then, when dealing with such speeds, to calculate simply the skin resistance, and approximate to the total resistance by adding a reasonable amount to allow for wave resistance. *Low Speeds.*

As the speed increases, this method becomes unsafe, and indeed is scarcely to be commended for any speed. But up to speeds for which the speed length ratio is not above 1.2, it is allowable to denote the wave resistance by bV^4. This gives us $R = fSV^{1.83} + bV^4$. *Practical Speeds.*

This is the practical working formula which I propose to adopt. The question of the value of b in a given case will be dealt with later.

At speeds for which the speed length ratio is greater than 1.2, we are driven to the complete formula,

$$R = fSV^{1.83} + V^4\left(A^2 + B^2 + 2kAB \cos\frac{m}{c^2} 646°\right).$$

The values of A, B, and k could, as things stand at present, be determined closely only by model experiments, and any one so situated as to be able to make model experiments need not struggle with approximate formulæ, but can adopt Froude's Method *in toto*. Fortunately for the vast majority who have not access to model tanks, speeds for which the speed length ratio is greater than 1.2 are exceptional, having been reached only in one or two instances by vessels other than torpedo boats. *Extreme Speeds.*

CHAPTER II.

THE PROPELLER.

§ 1. *Preliminary and Definitions.*

Propeller considered apart from Ship.
In discussing resistance I considered the ship alone — apart from its means of propulsion. I propose in the present chapter to discuss the propeller alone, — apart from the ship, — taking up later the modifications in the action of the propeller due to its connection with the ship. The problems to be solved are simplified by attacking them in detail.

Theories in Two Classes
There have been innumerable theories of the action of the propeller. They can be divided into two classes. To the first class may be assigned the theories, which consider the effect of the propeller upon the water, and from the motion of the water deduce the reaction upon the propeller. The "disc theory," so called, of Rankine is a notable example of the first class.

To the second class belong the theories which consider only the action of the water upon the propeller, using semi-empirical or experimental methods for dealing with it. The "blade theory" of Froude is a type of this class.

Disc Theory.
At first sight the first class of theories would seem to have the advantage. Given a certain amount of water having a certain change of velocity impressed upon it, the reaction resulting can apparently be calculated at once from the known density of water. This method would possess a beautiful simplicity if we knew the exact effect of a propeller upon the water which it passes through, and if the propeller blades were frictionless. Rankine assumed that a

§ 1.   PRELIMINARY AND DEFINITIONS.   63

screw propeller gave to a column of water, having a sectional area equal to the "disc" swept by the propeller, a sternward velocity corresponding to the slip. That this is impossible is manifest to any one who has seen a propeller as ordinarily fitted, and considered its working. Useful results may be obtained by supposing only a certain fraction of the disc area column to be given the sternward velocity of the slip. That fraction, however, can be determined only by experiment or semi-empirical methods. The friction of the propeller will still remain to be dealt with.

A curious feature about most of the theories of the first class is that they consider only the change of velocity of the water acted on, while the change of pressure is a matter of equal importance.

It is for this reason that theorists of the first class so often give the maximum theoretical efficiency of a propeller as $1 - \left(\frac{\text{slip}}{2}\right)$ instead of $(1 - \text{slip})$, the true theoretical maximum, when neglecting friction.

Rankine gave the value $(1 - \text{slip})$ for ordinary propellers, but concluded that a form of propeller which would work without "shock" was capable theoretically of an efficiency $\left(1 - \frac{\text{slip}}{2}\right)$. How a propeller working in "solid water" could administer to it this mysterious "shock" Rankine did not explain.

In Froude's theory, typical of the second class, the face of a propeller blade is treated as if it were made up of a number of small inclined planes advancing through the water.   *Blade Theory.*

This theory appears logical, and is, I believe, the most simple in practical application. The elder Froude in bringing forward his theory before the Institution of Naval Architects, in 1876, confined himself principally to the discussion of a single small plane element of the face of a blade. From this and other causes the idea arose that Froude's theory, followed out, would give results at variance with common and good practice. I hope to demonstrate that this is not the case. Before proceeding further with the subject, however, it will be well to give some brief definitions and state the notation that will be used.

Right and Left handed Propellers.   Figure 19 shows a four-bladed right-handed propeller. A right-handed propeller, viewed from aft, turns with the hands of a watch when driving its ship ahead. Under similar circumstances a left-handed propeller turns against the hands of a watch.

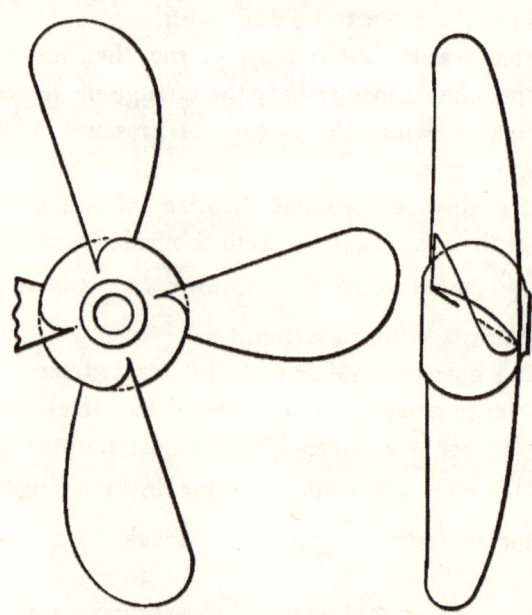

Fig. 19. — FOUR-BLADED RIGHT-HANDED SCREW PROPELLER.

Face.   The "face" of a blade is the rear face, the side which drives the water aft when the ship is going ahead.

Back.   The "back" of a blade is the side opposite the face. Care must be taken to avoid confusion, from the fact that the face of a blade is aft, and the back is forward.

Leading Edge. Following Edge.   The "leading edge" of a blade is the edge which leads; *i.e.* cuts the water first when the screw is turning ahead. The "following edge" is opposite the leading edge.

§ 1. PRELIMINARY AND DEFINITIONS. 65

The "diameter" of a screw is the diameter of the circle described by the tips of the blades. In symmetrical two and four bladed screws it is simply the distance from the "tip" or outermost part of one blade to that of the opposite blade. *Diameter. Tip.*

The "pitch," at a given point of the face, is the distance in the direction of the axis of the shaft which an elementary area of the face at the point, if attached by a rigid radius to the axis, would move during one revolution, if working in a solid fixed nut. *Pitch.*

The pitch may be different at every point of the face. If it is the same at all points, we say the pitch is "uniform."

If it is greater along the following than the leading edge, we say the pitch "increases axially." If it grows greater as we leave the centre, we say the pitch "increases radially." The term "increasing pitch," used without qualification, always refers to axially increasing pitch. *Increasing Pitch.*

The "area" or "developed area" of a blade is the surface of its face, and the "blade area" of a screw, sometimes called its "helicoidal area," is the area of all its blades. The "disc area" of a screw is the area of the circle described by the tips of its blades. *Blade Area. Disc Area.*

The "boss" or "hub" of a screw is the cylindrical or spherical centre to which the blades are attached, and which takes hold of the shaft. *Boss.*

The "pitch angle" at any point of a screw is the angle between a tangent plane at that point and a thwartship plane. *Pitch Angle.*

Thus, Figure 20, if LNP is a section of the tangent plane at O, OD being the fore and aft line, and OM a section of the thwartship plane, the angle POM is the pitch angle.

When a propeller is working with "slip," it advances during each revolution a distance less than the pitch, the difference between its actual advance and the pitch indicating the amount of "slip." When working with slip, an element such as LL, in Figure 20, does not advance parallel to itself along *Slip. Slip Angle.*

OP, but along a line such as *OS*, inclined to *OP*. The angle *SOP* is called the "slip angle."

Pitch and Diameter Ratios.
The "pitch ratio" of a screw is the ratio Pitch ÷ Diameter. Conversely, the diameter ratio is the ratio Diameter ÷ Pitch.

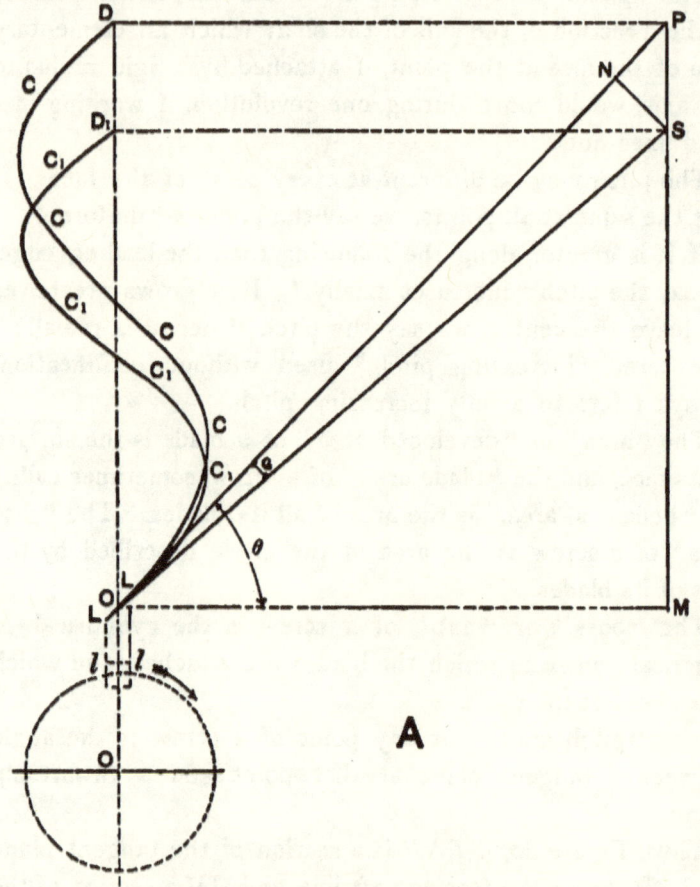

Fig. 20. — MOTION OF ELEMENT OF SCREW FACE.

It is necessary in dealing with screws to consider the whole blade from the tip in. Now taking pitch constant, as we come in toward the centre, the pitch ratio increases inversely with the diameter, while the diameter ratio decreases directly as the diameter. Consider a screw of 8 feet diameter and 10 feet pitch, and suppose the hub 1 foot in diameter.

Then the pitch ratio ranges from 1.25 at the tips to 10 at the hub.

The diameter ratio ranges from .8 to .1.

Used without qualification, pitch ratio and diameter ratio usually refer to the extreme diameter of the screw.

In what follows —

d denotes diameter in feet;
r denotes radius in feet;
p denotes pitch in feet;
A denotes surface in square feet;
R denotes revolutions per minute;
V denotes speed of ship in knots;
V' denotes speed of propeller in knots.

Symbols.

By "speed of propeller" is meant the speed corresponding to no slip; thus if speed were measured in feet per minute, the speed of the propeller would always be denoted by pR.

s denotes slip expressed as a fraction of the speed of the propeller; *i.e.*

$$s = \frac{V' - V}{V'}$$

y denotes diameter; ratio $= \frac{d}{p}$.

§ 2. *Element of Face of Blade treated as a Plane.*

I propose now to discuss the action of a small isolated plane area when it is given a motion similar to that of a small element of the face of a propeller blade.

Referring to Figure 20, let LL denote a small plane set at the angle θ with an axis O which revolves while it advances.

Then supposing first that there is no slip, during one revolution, LL, moving along a spiral path shown in plan by

Nature of Motion of Plane Element.

$OCCD$, will advance to D. If slip is present so that the advance instead of being OD is some lesser quantity, say OD_1, the spiral path of advance will be denoted in plan by $OC_1C_1D_1$.

Let r denote the length of the radius from the centre O to middle of the element LL.

Then lay off $OM = 2\pi r$ = circumference of circle described by LL when looked at from aft. Set up $MP = OD$ = the pitch of the element. Let $MS = OD_1$ = advance in one revolution when there is slip. Evidently $MS = MP (1 - \text{slip})$.

Let R denote the revolutions per minute of the element. Then OSM is a diagram of velocities. For $OM \times R$ = circumferential velocity of the element, and $MS \times R$ = velocity in the direction of advance. Then $OS \times R$ denotes the actual velocity of the element which is in the direction OS.

Denote POS, the slip angle, by ϕ, and the area of the small element LL by dA.

Reactions on Element.
Let dP denote the normal and dF the tangential reaction upon LL.

Then using Froude's formula for planes at small inclinations, $dP = a \cdot dA \cdot (R \times OS)^2 \sin\phi$, where a is a constant determined by experiment. A convenient name for a is the "Thrust Constant." As previously stated, its value as determined by Froude, for pressures in pounds and velocity in feet per second, is 1.7. The number expressing its value changes of course if the units of force and velocity are changed.

When we come to express dF, the friction upon the element, we are met by the fact that it is difficult to determine the exact velocity of the water over the surface. While the speed of the element is denoted by OS, it is moving through water which it has set in motion more or less. It appears certain that the true velocity of gliding of the water should be expressed by a quantity greater than OS and less than OP. To be on the safe side and to facilitate calculations, I shall choose OP to denote the velocity of gliding.

§ 2.   ELEMENT OF FACE OF BLADE AS A PLANE.   69

The error involved in this must be small in any practical case, and negligible for those parts of an actual propeller blade where the friction is greatest.

Then if f denote the coefficient of friction of the plane element, $dF = f \cdot dA \cdot (R \times OP)^2$.

Referring now to Figure 21, where LL is shown on a larger scale than in Figure 20, let dT denote the thrust due to LL, and dM the tangential force. dT will be equal and opposite *Thrust and Turning Force.*

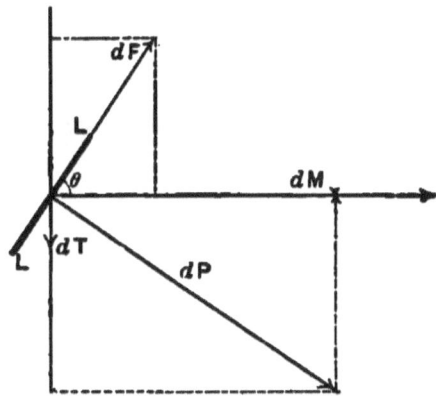

Fig. 21. — TO ACCOMPANY FIG. 20.

to the resistance which the element may be supposed to be overcoming, and dM must balance the turning force being exerted upon the element.

By resolution of forces as shown in the figure,

$$dT = dP \cos\theta - dF \sin\theta\ ;$$
$$dM = dP \sin\theta + dF \cos\theta.$$

Returning now to Figure 20, let

$$PM = p,\quad OM = 2\pi r = \pi d,\quad SM = p(1-s),\text{ and } \frac{d}{p} = y.$$

Geometry of the Motion.

Draw SN perpendicular to OP.

Then

$$(OS)^2 = (OM)^2 + (MS)^2 = \pi^2 d^2 + p^2(1-s)^2$$
$$= p^2\{\pi^2 y^2 + (1-s)^2\}\text{ whence } OS = p\sqrt{\pi^2 y^2 + (1-s)^2}.$$

$$\sin \theta = \frac{PM}{OP} = \frac{p}{\sqrt{p^2 + \pi^2 d^2}} = \frac{1}{\sqrt{1 + \pi^2 y^2}},$$

$$\cos \theta = \frac{OM}{OP} = \frac{\pi d}{\sqrt{p^2 + \pi^2 d^2}} = \frac{\pi y}{\sqrt{1 + \pi^2 y^2}},$$

$$\sin \phi = \frac{SN}{OS} = \frac{PS \cos \theta}{p\sqrt{\pi^2 y^2 + (1-s)^2}}$$

$$= \frac{p \cdot s \cdot \pi y}{p\sqrt{\pi^2 y^2 + (1-s)^2}\sqrt{1 + \pi^2 y^2}}$$

$$= s \frac{\pi y}{\sqrt{1 + \pi^2 y^2}\sqrt{\pi^2 y^2 + (1-s)^2}},$$

$$(OP)^2 = (OM)^2 + (PM)^2 = p^2 + \pi^2 d^2 = p^2(1 + \pi^2 y^2).$$

Thrust. I am now in a position to express dT and dM in different forms.

$$dT = dP \cos \theta - dF \sin \theta = adA(R \times OS)^2 \sin \phi \cos \theta$$
$$- fdA(R \times OP)^2 \sin \theta$$

$$= dA \times R^2 \left[\frac{ap^2\{\pi^2 y^2 + (1-s)^2\} s \cdot \pi y \cdot \pi y}{\sqrt{1 + \pi^2 y^2}\sqrt{\pi^2 y^2 + (1-s)^2}\sqrt{1 + \pi^2 y^2}} \right.$$
$$\left. - \frac{fp^2(1 + \pi^2 y^2)}{\sqrt{1 + \pi^2 y^2}} \right]$$

$$= p^2 \cdot R^2 \cdot dA \left[a \cdot s \cdot \frac{\pi^2 y^2 \sqrt{\pi^2 y^2 + (1-s)^2}}{1 + \pi^2 y^2} \right.$$
$$\left. - f\sqrt{1 + \pi^2 y^2} \right].$$

Also, $dM = dP \sin \theta + dF \cos \theta$

$$= a.R^2 dA (OS)^2 \sin \phi \sin \theta + fR^2 dA (OP)^2 \cos \theta$$

$$= R^2 dA \left[\frac{a \cdot p^2\{\pi^2 y^2 + (1-s)^2\} s \cdot \pi y}{\sqrt{1 + \pi^2 y^2}\sqrt{\pi^2 y^2 + (1-s)^2}\sqrt{1 + \pi^2 y^2}} \right.$$
$$\left. + \frac{fp^2(1 + \pi^2 y^2)\pi y}{\sqrt{1 + \pi^2 y^2}} \right]$$

$$= p^2 R^2 dA \left[as \frac{\pi y \sqrt{\pi^2 y^2 + (1-s)^2}}{1 + \pi^2 y^2} + f\pi y \sqrt{1 + \pi^2 y^2} \right].$$

§ 2. ELEMENT OF FACE OF BLADE AS A PLANE.

Determine now the useful work, *i.e.* the work done in the direction of thrust, denoted by dU, and the gross work delivered to the element, denoted by dN. *Useful Gross Work.*

Evidently (Figures 20 and 21)

$$dU = dT \times MS = dT \times pR(1-s)$$

$$= p^3 R^3 dA \left[as(1-s)\frac{\pi^2 y^2 \sqrt{\pi^2 y^2 + (1-s)^2}}{1+\pi^2 y^2} - f(1-s)\sqrt{1^2 + \pi^2 y} \right]$$

$$dN = dM \times \pi d \times R = dM \cdot \pi y \cdot pR$$

$$= p^3 R^3 dA \left[\frac{as\pi^2 y^2 \sqrt{\pi^2 y^2 + (1-s)^2}}{1+\pi^2 y^2} + f \cdot \pi^2 y^2 \sqrt{1+\pi^2 y^2} \right]$$

There is no good result obtained by writing these complicated expressions at full length. So denote

$$\frac{\pi^2 y^2 \sqrt{\pi^2 y^2 + (1-s)^2}}{1+\pi^2 y^2} \text{ by } X^1;$$

$$\sqrt{1+\pi^2 y^2} \text{ by } Y;$$

$$\pi^2 y^2 \sqrt{1+\pi^2 y^2} \text{ by } Z.$$

Then
$$dU = p^3 R^3 (1-s)(asX^1 - fY)dA,$$

$$dN = p^3 R^3 (asX + fZ)dA.$$

The efficiency of the element, denoted by e, is the ratio between the useful work done by it and the gross work delivered to it. *Efficiency.*

$$\text{Then } e = \frac{dU}{dN} = (1-s)\frac{asX_1 - fY}{asX^1 + fZ}.$$

The quantities X^1, Y, and Z can be readily calculated. *X^1, Y, and Z.*

The first, X^1, depends upon the diameter ratio y, and the slip s, while Y and Z involve diameter ratio and known constants only.

Table VII. gives values of X^1 for various diameter ratios and slips, and Table VIII. gives values of Y and Z for various diameter ratios.

Curves of Efficiency.

It is to be observed that the value of the efficiency depends upon the relative value of a and f, the thrust constant and the coefficient of friction. The expression for the efficiency may be written

$$e = (1-s)\frac{\frac{a}{f}sX^1 - Y}{\frac{a}{f}sX_1 + Z}.$$

Froude, as a result of experiments made before 1876, assigned to a the value 1.7, and to f the value .0085, the unit of speed being one foot per second, and the unit of force one pound.

The value of f is double what may be called its natural value, to allow us to deal with the area of the face of a blade only, and yet to express the friction of the equal area of the back.

For $a = 1.7$ and $f = .0085$ $\frac{a}{f} = 200$.

The expression for efficiency then becomes

$$e = (1-s)\frac{200\,sX^1 - Y}{200\,sX^1 + Z}.$$

Figure 22 shows in full lines the curves of efficiency of elementary areas of the diameter ratios indicated, plotted on slips as abscissæ. They were obtained from the above formula by the use of Tables VII. and VIII. The dotted curves in Figure 22 show very closely the efficiency curves of actual small screw propellers — the data being taken from papers read by the elder and younger Froude before the Institution of Naval Architects in 1883 and 1886 respectively.

It seems impossible for any one to observe the essentially similar nature of the full and dotted curves of Figure 22 without feeling the force of the words with which the elder Froude ended his paper on the subject before the Institution of Naval Architects in 1878: "No theoretical treatment of

§ 2.   ELEMENT OF FACE OF BLADE AS A PLANE.

the action of an actual screw can be sound which does not incorporate and mainly rest on the principles embodied in the treatment of the problem of the plane, and indeed the character of the results must, in their most essential features, be the same in both cases."

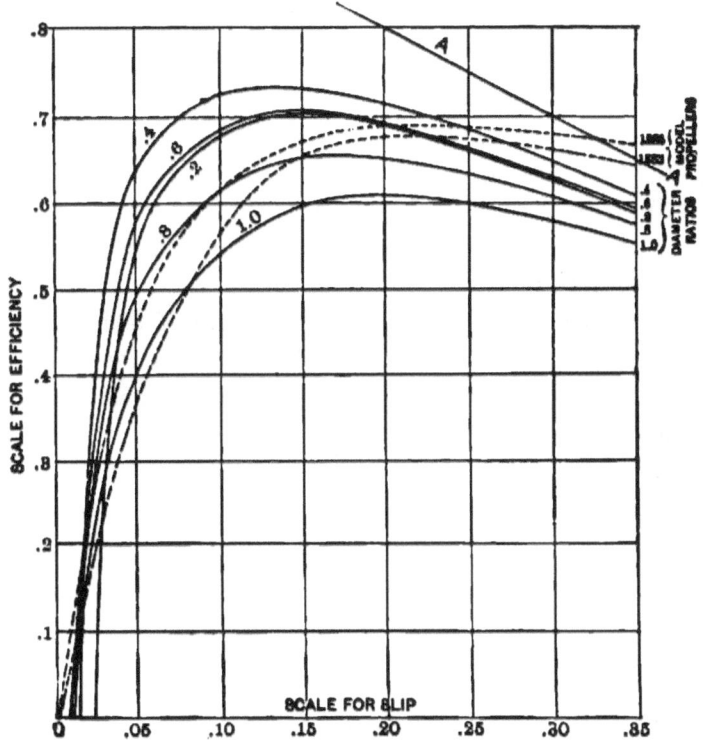

Fig. 22. — EFFICIENCY OF ELEMENTS AND OF MODEL PROPELLERS.

To determine X^1, we have the expression

$$X_1 = \frac{\pi^2 y^2 \sqrt{\pi^2 y^2 + (1-s)^2}}{1 + \pi^2 y^2}.$$

Substitution of X for X^1.

A little reflection upon this formula and an inspection of Table VII. will make it evident that the difficulties of tabulating and handling X^1 are largely increased by the presence of $(1-s)^2$ under the radical sign in the numerator.

Now the slips with which propellers work in practice vary from .15 to .30, as a rule. I have accordingly inserted in Table VIII. a column giving X, which is the value of X^1, for a slip of .2. Table IX. gives comparative efficiencies for elements of various diameter ratios obtained by using X^1 and X. It is seen that for slips occurring in practice the efficiencies are practically identical.

The useful and gross work obtained by using X will be slightly less than if X^1 were used for slips below .2, and slightly greater for slips above .2. We shall see later, however, that we thus probably obtained a closer approximation to the working of actual propellers.

So henceforth I shall discard X^1 and use X, as being simpler and better in every way. Then the quantities X, Y, and Z of Table VIII. are functions only of the diameter ratio and determinate constants.

Maximum Efficiency of Element. In concluding the discussion of the plane element, I propose to deal with its maximum efficiency under various conditions.

Our expression for efficiency is

$$e = (1-s)\frac{s\frac{a}{f}X - Y}{s\frac{a}{f}X + Z}.$$

Denote $\frac{a}{f}X$ by c;

Then $e = (1-s)\dfrac{cs - Y}{cs + Z}$,

or $\qquad e(cs + Z) = cs - Y - cs^2 + sY.$ \hfill (1)

Differentiating with respect to s,

$$\frac{de}{ds}(cs + Z) + ce = c - 2cs + Y. \qquad (2)$$

§ 2.   ELEMENT OF FACE OF BLADE AS A PLANE.

At the slip corresponding to maximum efficiency, the efficiency curve is horizontal, or $\frac{de}{ds}=0$.

Then if e_m and s_m denote maximum efficiency and slip corresponding, we have from (2)

$$ce_m = c - 2cs_m + Y, \qquad (3)$$

and from (1)

$$e_m(cs_m + Z) = cs_m - Y - cs_m^2 + s_m Y. \qquad (4)$$

Solving (3) and (4) for e_m and s_m, we have

$$e_m = \frac{1}{c}\{\sqrt{Z+c} - \sqrt{Z+Y}\}^2,$$

$$s_m = \frac{1}{c}\{-Z + \sqrt{(Z+c)(Z+Y)}\}.$$

Also if ϕ_m denote the angle of slip corresponding to s_m,

$$\sin\phi_m = s_m \frac{\pi y}{\sqrt{1+\pi^2 y^2}\sqrt{\pi^2 y^2 + .64}}.$$

Figure 23 shows graphically the maximum efficiencies attainable by elements of various diameter ratios on three suppositions as to $a+f$. The slip angles corresponding are also shown. It is to be observed that the absolute maximum efficiency in each case is found at about a diameter ratio $=.35$.

This applies only to a single element and does not indicate that an actual propeller will develop maximum efficiency if its diameter ratio (corresponding to the extreme diameter) is .35.

Figure 23 if rightly interpreted affords conclusive evidence that the diameter ratio of the actual propeller must be greater than .35 for maximum efficiency. For suppose it were but .35. Then the tips of the blades would be working with high efficiency, but the bulk of the blade (within the tips, where the diameter ratio would be less than .35) would be working with much less efficiency.

The extreme diameter ratio should evidently be such that the bulk of the work is done under the most efficient conditions.

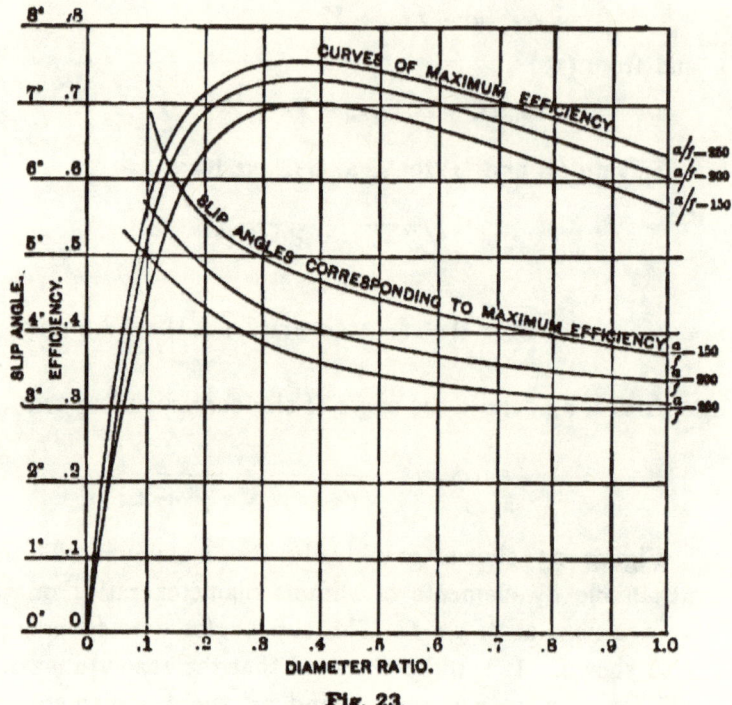

Fig. 23.

§ 3. *Extension of Formula for Plane to Propeller.*

Difference between Small Plane and Element of Face of Propeller.

The results obtained so far apply, it must be remembered, to an imaginary isolated small plane given the same motion as an elementary area on the face of an actual propeller blade.

What are the points of difference between the small plane and the element of a propeller blade face?

1. The plane is isolated; the blade element is surrounded by other elements.

2. The plane is of negligible thickness; the blade element has the thickness of the blade behind it.

3. The plane is supposed to move in still water; the blade element moves in water already set in motion by the action of the propeller.

The exact effect of the differences enumerated above cannot be predicated on theoretical grounds. It appears probable, however, that their effect is not great, and it is exceedingly probable *a priori* that it is simply equivalent to a modification of the thrust constant a and the coefficient of friction f, so that results obtained from experiments on planes will not be exactly applicable to actual propellers.

Without taking up at present the question of the exact values of a and f for a given case, let us proceed to extend the methods already explained to the full-sized blade.

I shall take a blade of uniform pitch.

Figure 24 shows an expanded propeller blade. To properly develop a blade, it should be first expanded or twisted until it is all in one plane, and then a new contour line drawn through the extremities of lines perpendicular to the centre line of the blade and of the length of corresponding circular arcs to the contour of the twisted blade. To deduce a blade from a given developed shape, the above process should be reversed. Referring to Figure 24, let l denote the length of the strip at radius r, as shown, the width of the strip being dr. Then ldr is an elementary area corresponding to dA in the formula for the plane.

Developed or Expanded Blade.

Formula for Single Blade.

Then for the elementary strip,

$$dU = p^3 R^3 l dr (1-s)(asX - fY),$$

$$dN = p^3 R^3 l dr (asX + fZ).$$

Let U and N denote the useful and gross work of the whole blade.

Then
$$U = \int dU = p^3 R^3 \int (1-s)(asX - fY) l \cdot dr;$$
$$N = \int dN = p^3 R^3 \int (asX + fZ) l \cdot dr.$$

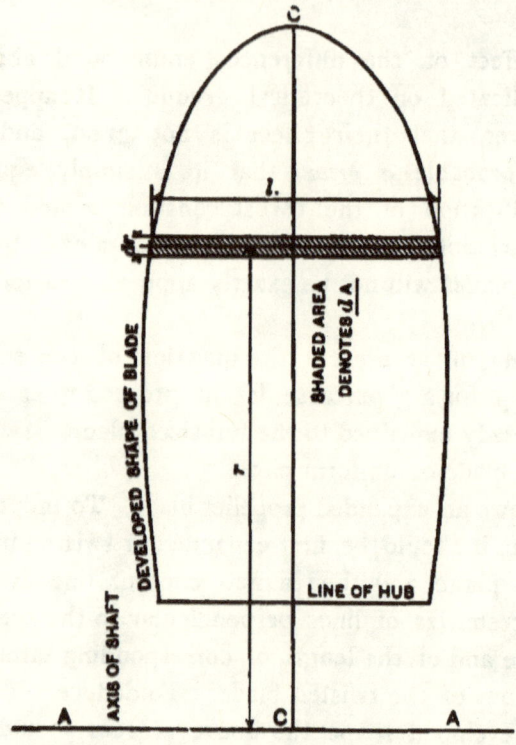

Fig. 24. — EXPANDED PROPELLER BLADE.

It is convenient to replace l and dr by the ratios $\dfrac{l}{d}$ and $\dfrac{dr}{d}$, d denoting the extreme diameter of the propeller. Making these substitutions and noting that s is constant for the assumed uniform pitch, we may rewrite our expressions,

$$U = p^3 R^3 d^2 \left[as(1-s) \int X \frac{l}{d} \cdot \frac{dr}{d} - f(1-s) \int Y \frac{l}{d} \cdot \frac{dr}{d} \right],$$
$$N = p^3 R^3 d^2 \left[as \int X \frac{l}{d} \frac{dr}{d} + f \int Z \frac{l}{d} \frac{dr}{d} \right],$$

the integration in each case extending over the whole blade.

§ 3. EXTENSION OF FORMULA

Now the quantities

The Characteristics.

$$\int X \frac{l}{d} \frac{dr}{d}, \int Y \frac{l}{d} \frac{dr}{d} \text{ and } \int Z \frac{l}{d} \frac{dr}{d}$$

depend on the diameter ratio, the proportions, and the shape of the blade, but not on the size. Their value in a given case may be said to be "characteristic" of the blade. Then denote $\int X \frac{l}{d} \frac{dr}{d}$ by X, called the X characteristic;

$\int Y \frac{l}{d} \frac{dr}{d}$ by Y, called the Y characteristic;

and $\int Z \frac{l}{d} \frac{dr}{d}$ by Z_e, called the Z characteristic.

We have then

$$U = p^3 R^3 d^2 [as(1-s) X_e - f(1-s) Y_e],$$
$$N = p^3 R^3 d^2 [as X_e + f Z_e].$$

As will be seen later, the calculation of X_e, Y_e, and Z_e from the developed plan of a blade is a simple process.

So far I have considered only a single blade, and taken no account of the units of speed and power used.

The numerical expressions for the values of a and f will depend upon these units. Now it is convenient to retain nearly the same numerical values for a and f as if the knot were the unit of speed, and the pound the unit of force.

Number of Blades and Units used.

But we have introduced horse-power, and it is customary and convenient to express pitch and diameter in feet, and revolutions by the number made in one minute. Evidently, then, a factor must be introduced.

Expressing pitch in feet and revolutions by the number made in one minute, the unit of speed is one foot per minute. Now one foot per minute = $\frac{80}{8080}$ knots. Hence to denote thrust or force in pounds we should multiply the expressions above by $(\frac{80}{8080})^2$.

Now thrust $\times pR =$ work done in foot-pounds per minute $= 33{,}000 \times$ the horse-power developed.

Then the factor to be introduced is

$$\tfrac{1}{33000} \times (\tfrac{6080}{6080})^2 = \tfrac{1}{338858666}.$$

This is an unwieldy quantity. If for 338858666 we write 333333333⅓, our results will be much simplified without any real error. The only effect will be that a and f will be expressed by numbers about one and a half per cent smaller than if the change had not been made.

The quantity $(pR)^3$ appearing in our expressions will be an unwieldy one in practical cases, since pR is seldom less than 1000 or greater than 3000.

So let us substitute for $(pR)^3$ the equal quantity

$$1000000000 \left(\frac{pR}{1000}\right)^3.$$

If n denote the number of blades of a propeller, its thrust, horse-power, etc., will be n times those of a single one of its blades.

Formulæ for Complete Propeller.

So we have finally for a propeller

$$U = \frac{1000000000}{333333333\tfrac{1}{3}} n \left(\frac{pR}{1000}\right)^3 d^2 [as(1-s)X_e - f(1-s)Y_e]$$

$$= 3 n \left(\frac{pR}{1000}\right)^3 d^2 [as(1-s)X_e - f(1-s)Y_e].$$

Similarly, $\quad N = 3 n \left(\dfrac{pR}{1000}\right)^3 d^2 [asX_e + fZ_e].$

These expressions are rigorously deduced from the fundamental assumptions which were discussed and justified as they were made.

§ 4. *Values of a, f, and the Characteristics.*

Method of Determining a and f.

The question now arises, how are we to assign values to the constants a and f for a given propeller?

Suppose we determined by experiment, with a known propeller, working at a known number of revolutions with

§ 4. VALUES OF a, f, AND THE CHARACTERISTICS. 81

a known slip, the values of U and N. We should then have two equations to determine the values of a and f. Determination of U and N for the same propeller, at a different number of revolutions, and working with a different slip, would enable us to determine a second set of values of a and f. So by sufficiently extending our experiments we could determine values of a and f for the propeller in hand throughout the range of slip likely to occur in practice.

There have never been such experiments made, so far as I am aware, upon propellers of any size unattached to an actual vessel. The most valuable experimental results available will be found published in a paper read by Mr. R. E. Froude before the Institution of Naval Architects in 1886. The experiments were made upon four-bladed model propellers 0.68 of a foot in diameter. *Froude's Experimental Propeller.*

The blades were elliptical in shape, and each blade developed would have been (but for the boss) a perfect ellipse touching the axis of the shaft. The extreme width, or minor axis of the ellipse, was equal to .4 of the radius.

The elliptical form was departed from slightly in the vicinity of the boss, which was in diameter about $\frac{1}{10}$ the diameter of the propeller.

Figure 25 shows very approximately Froude's elliptical blade twisted into one plane. Before proceeding further I shall show in detail the calculation of X_e, Y_e, and Z_e in this case. For the blade shown the extreme diameter ratio is .8, corresponding to a pitch ratio of 1.25. The radius from the tip to the boss is divided into equal parts at points corresponding to diameter ratio .1, .2, .3, and so on. At each point of division is determined the value of l, the width of the blade, and the value of $\frac{l}{d}$. The diameter d is of course the extreme diameter, not the diameter to the point where the width is measured. *Characteristics of Froude's Blade.*

From Table VIII. the values of X corresponding to the

diameter ratios at the several points of division are picked out and multiplied into the corresponding values of $\frac{l}{d}$.

This gives the data necessary to enable us to draw a curve of $X\frac{l}{d}$, as shown in Figure 25, upon the radius, as base.

The area of this curve is readily determined, preferably by an ordinary planimeter.

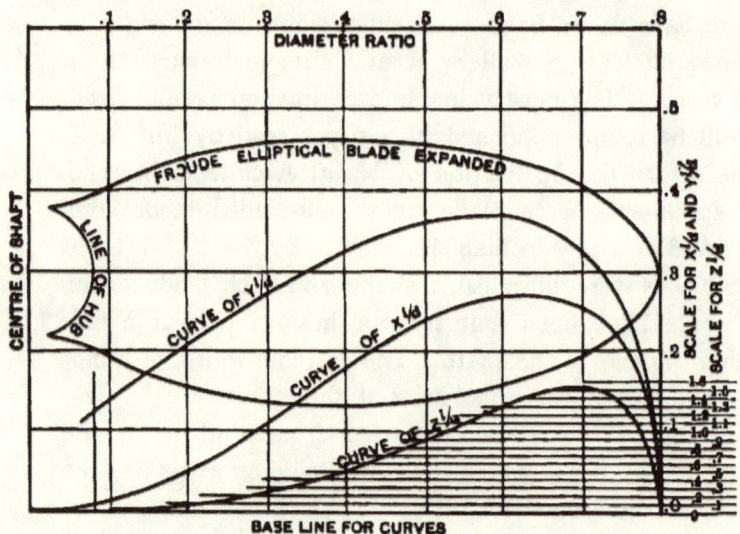

Fig. 25.—SHAPE AND CHARACTERISTICS OF FROUDE EXPERIMENTAL BLADE.

The value of X_e bears a ratio to the area of the curve of $X\frac{l}{d}$, depending upon the scales used.

Thus if A_x denote the area of the curve in square inches, while the curve was drawn on the scale 1 inch $= n$, and the blade diagram on the scale 1 inch $= g$ feet $= \frac{1}{m}$ of the radius, we shall have $X_e = \frac{n}{2m} A_x$.

The curves of $Y\frac{l}{d}$ and $Z\frac{l}{d}$ are also shown in Figure 25, and the values of Y_e and Z_e are readily determined from

§ 4. VALUES OF a, f, AND THE CHARACTERISTICS. 83

them, following exactly the same process as in determining X_c.

The values of the characteristics deduced from Figure 25 are

$$X_c = .075,$$
$$Y_c = .127,$$
$$Z_c = .322.$$

These are for the Froude elliptical blade of diameter ratio $= .8$. The characteristics of the same shape of blade

Curves of Characteristics.

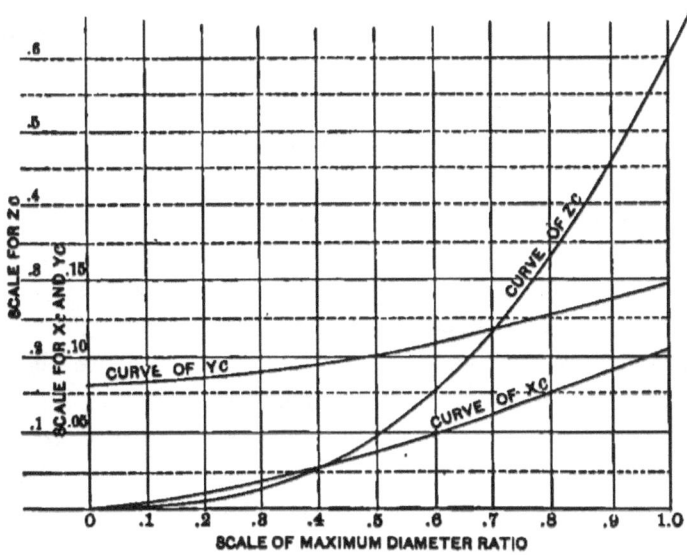

Fig. 26. — FROUDE EXPERIMENTAL PROPELLER. CURVES OF CHARACTERISTICS.

with other diameter ratios may be readily calculated, and from a sufficient number of such determinations curves may be drawn, showing the three characteristics plotted on the values of maximum diameter ratio as abscissæ.

Figure 26 shows the characteristics of Froude's elliptical blade up to a diameter ratio of 1.0.

Constants of Froude's Propeller.

Having determined the characteristics of the blade used in Froude's experiments, we can from the curves of useful horsepower and efficiency published in the I. N. A. paper of 1886, deduce values of a and f in the manner outlined in the first part of this section.

The curves from the experimental results, as given by Froude, are not in the shape best for the determination of a and f. It is found, however, that the values of a and f deduced from them are fairly consistent.

Value of f.

The average value of f is about .045. This corresponds to a coefficient of friction of .0225, which, while double that for a smooth plane surface, seems not unreasonable for a propeller blade possessing a certain thickness and edge resistance.

Value of a.

Upon attempting to determine a, it was found to vary with the diameter ratio of the propeller.

This is entirely natural; for the less the diameter ratio, the thicker the slices of water between successive blades, and the less the interference of the blades with each other.

The value of a depends also upon the number of blades. Froude states that the thrusts of four, three, and two bladed propellers of identical blades are as 1.000 : .865 : .65 instead of as 1.000 : .75 : .50, as would be the case if a were constant. The increase of the thrust per blade as the number of blades is diminished is probably due to the diminution of interference with the fewer blades.

The values of a as finally determined from the analysis of Froude's results may be expressed as below. If m denote the extreme diameter ratio —

For four-bladed propellers, $a = 8.4 - 1.0\,m$.

For three-bladed propellers, $a = 9.4 - 1.2\,m$.

For two-bladed propellers, $a = 10.4 - 1.4\,m$.

While these values of a apply directly only to the experimental propeller used by Froude, they appear to be close approximations to the exact values for full-sized composition or manganese bronze propellers, such as are ordinarily fitted to men-of-war and fast passenger steamers.

For the thicker and blunter cast iron blades customary in the merchant marine the values of a are less than those given by the above. I shall say more about the values of a and f in a later chapter.

In any propeller the "mean width of blade" Width Ratio.

$$= \frac{\text{area of blade}}{\text{radius of blade} - \text{radius of boss}}$$

For the experimental propeller just discussed, the ratio between the mean width of blade and the diameter of the propeller was .166. This ratio is a useful quantity, and it is well to give it a name. I shall call it the "mean width ratio." The mean width ratio of propellers as usually fitted is seldom so low as .166.

§ 5. *Efficiency and Power of Various Shapes of Blades.*

I propose in this section to discuss the effect of shape of blade upon the efficiency and power of a propeller. In Figure 27 are shown developed five shapes of blade, all of the same area, radius, and radius of boss. Their principal dimensions are given below. Data of Five Shapes.

Shape	No. 1.	No. 2.	No. 3.	No. 4.	No. 5.
Diameter of boss	$\tfrac{2}{6}d$	$\tfrac{2}{6}d$	$\tfrac{2}{6}d$	$\tfrac{2}{6}d$	$\tfrac{2}{6}d$
Maximum width	.2023 d	.1686 d	.2 d	.2023 d	.2023 d
Minimum width	.1349 d	.1686 d		.1349 d	.1349 d
Width ratio	.1686	.1686	.1686	.1686	.1686

It is seen that the five blades differ only in shape. No. 3 is identical with Froude's experimental blade except as to boss. The boss has been taken in each case as $\frac{2}{8}$ of the diameter. This is a fair average of the sizes of boss customary in good practice.

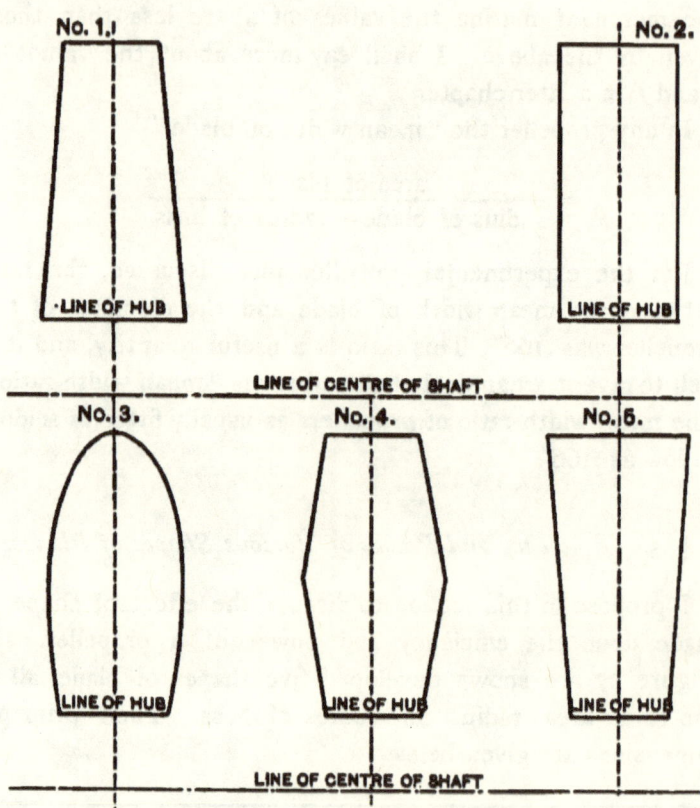

Fig. 27. — DEVELOPED SHAPES OF BLADE.

Characteristics. The X, Y, and Z characteristics of the five shapes have been calculated up to an extreme diameter ratio of 1.0, and are shown graphically by Figures 28, 29, and 30. It is to be observed that above diameter ratio of .5 blade No. 5 has always the greatest characteristic, and in each case the remainder follow in the order No. 2, No. 4, No. 1, No. 3.

§ 5.   EFFICIENCY AND POWER OF BLADES.

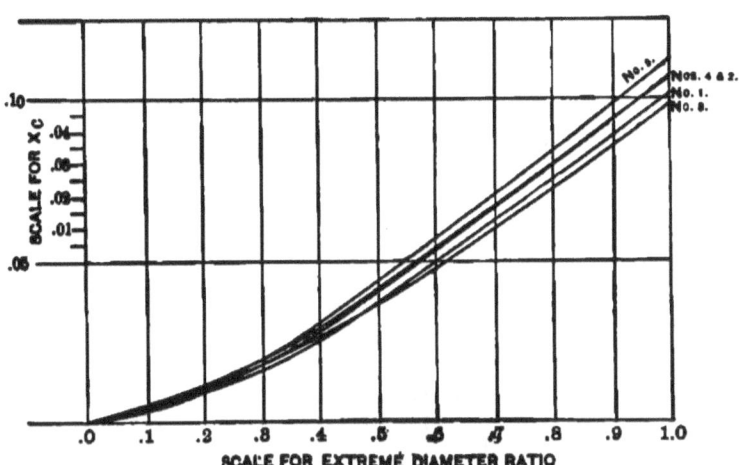

Fig. 28. — X CHARACTERISTICS OF FIVE SHAPES OF BLADE.

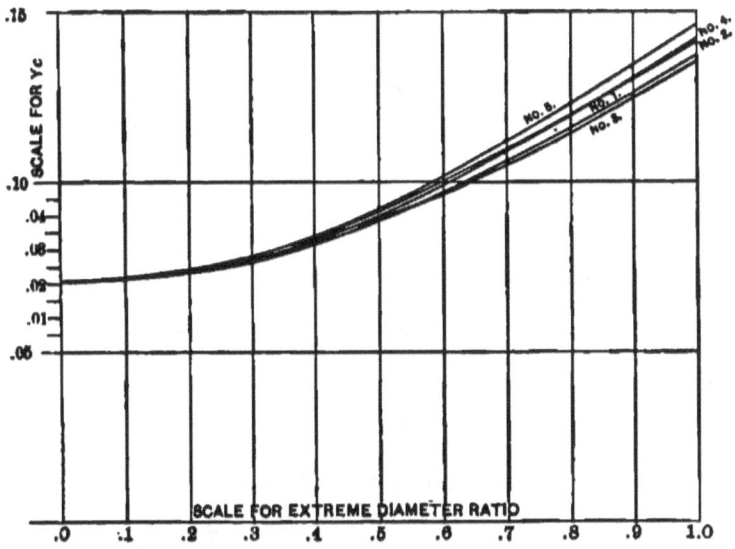

Fig. 29. — Y CHARACTERISTICS OF FIVE SHAPES OF BLADE.

88 RESISTANCE OF SHIPS. § 5.

Efficiency. Knowing the characteristics of a blade, the maximum efficiency and corresponding slip are readily obtained by the formulæ on page 75, substituting X_e, Y_e, and Z_e, for X, Y, and Z.

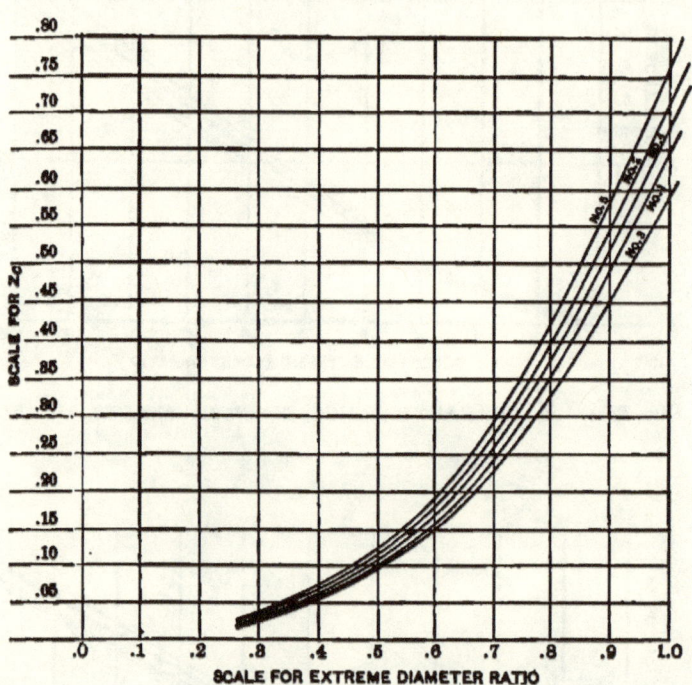

Fig. 30. — Z CHARACTERISTICS OF FIVE SHAPES OF BLADE.

We have for maximum efficiency,

$$e_m = \frac{1}{c}\{\sqrt{Z_e + c} - \sqrt{Z_e + Y_e}\}^2,$$

where $$c = \frac{a}{f} X_e,$$

and for slip corresponding to maximum efficiency,

$$s_m = \frac{1}{c}\{-Z_e + \sqrt{(Z_e + c)(Z_e + Y_e)}\}.$$

Assuming $\frac{a}{f} = 200$, the maximum efficiency of each shape of

§ 5. EFFICIENCY AND POWER OF BLADES.

blade for various diameter ratios was calculated with results tabulated below.

Maximum Diam. Ratio.	.4	.6	.8	1.0
		Maximum Efficiencies.		
Blade No. 1.	.713	.718	.698	.671
No. 2.	.714	.718	.697	.677
No. 3.	.717	.721	.704	.678
No. 4.	.717	.721	.706	.671
No. 5.	.718	.718	.695	.664
Average	.7158	.7192	.6988	.6722

While No. 3 appears to have slightly the advantage, the maximum efficiencies are practically identical, when the value of $\frac{a}{f}$ is the same for each shape. The same is true of the slips corresponding to maximum efficiency.

In Figure 31 I have plotted the average curve of maximum efficiency and of slip corresponding.

It does not appear, from the results just obtained, that there is any especial virtue in a particular form of blade. Common experience points to the same conclusion. Referring to Figure 31, it is seen that the absolute maximum efficiency is found at about the comparatively small diameter ratio of .5, but that the falling off of maximum efficiency is small until we reach a diameter ratio of .8 or so.

This agrees with R. E. Froude's observations. He stated that in his experiments, which extended from a diameter ratio of about .8 to one of about .5, there appeared to be a slight increase of maximum efficiency as the pitch ratio increased (*i.e.* as the diameter ratio diminished).

The agreement of Figure 31 with the results of common experience and of careful experiments tends to confirm the soundness of the assumptions from which it was deduced, and

Useful Power of Five Shapes.

to still further justify my adoption of the blade theory in treating the propeller.

To further fix our ideas and become familiar with the methods described, let us inquire into the useful horse-power delivered by five propellers with blades of the five shapes we have been discussing.

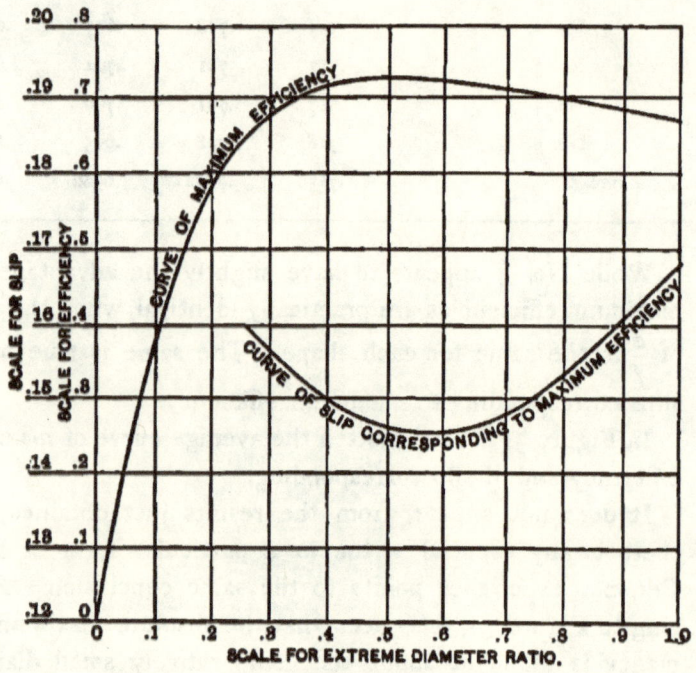

Fig. 31.—AVERAGE MAXIMUM EFFICIENCY AND SLIP FOR FIVE BLADES.

The expression for the useful horse-power U delivered by a propeller is (page 80)

$$U = 3n\left(\frac{pR}{1000}\right)^3 d^2[as(1-s)X_e - f(1-s)Y_e].$$

Take all five propellers as three-bladed 10 feet in diameter, of 14.286 feet pitch (diameter ratio = .7), and making 117.6 revolutions per minute.

§ 5   EFFICIENCY AND POWER OF BLADES.   91

Take for a and f their values as deduced from Froude's experiments, namely $f=.045$, $a=9.4-1.2\times.7=8.54$. Then our formula becomes

$$U = 3 \times 3 \frac{(14.286 \times 117.6)^2}{1000} \times (10)^2 (1-s)$$

$$\times \{8.56\, s X_e - .045\, Y_e\}$$

$$= 900 \times 4.741632 \times 8.56\, X_e (1-s) \left\{ s - \frac{.045\, Y_e}{8.56 X_e} \right\}$$

$$= 36530\, X_e (1-s) \left\{ s - .00526 \frac{Y_e}{X_e} \right\}.$$

The characteristics of the various blades as obtained from Figures 28, 29, and 30 for a diameter ratio of .7 are as below:

	X_e	Y_e	$Y_e + X_e$
No. 1.	.0630	.1088	1.726
No. 2.	.0670	.1118	1.668
No. 3.	.0613	.1080	1.763
No. 4.	.0663	.1108	1.672
No. 5.	.0700	.1143	1.632

Substituting the various characteristics in the general formula, we have

Results for Revolutions Constant.

For No. 1, $U = 2301\,(1-s)\,(s-.0091)$.
For No. 2, $U = 2448\,(1-s)\,(s-.0088)$.
For No. 3, $U = 2237\,(1-s)\,(s-.0093)$.
For No. 4, $U = 2420\,(1-s)\,(s-.0088)$.
For No. 5, $U = 2557\,(1-s)\,(s-.0086)$.

Figure 32 shows curves of useful horse-power plotted from the above five equations upon slips as abscissæ. The revolutions having been assumed constant at 117.6, increase of slip must be obtained by decreasing the speed of advance of the propeller.

Results for Speed of Advance Constant. Let us now determine similar curves when the speed of advance remains constant, and increase of slip is obtained by increasing the revolutions.

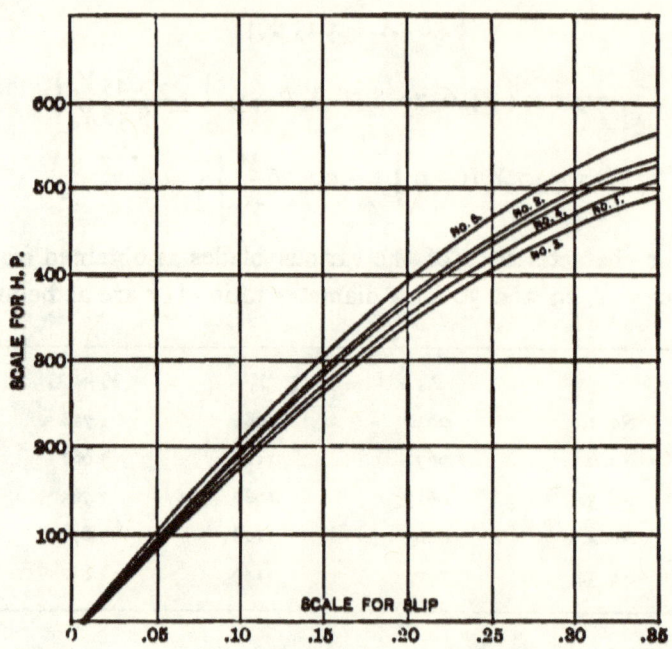

Fig. 32.—HORSE-POWER AT CONSTANT REVOLUTIONS.

Assume in the above case that the constant speed of advance is 1680 feet per minute, *i.e.* that at 117.6 revolutions the slip is nil. Then the only change necessary in the original formula is the substitution of $\frac{117.6}{1-s}$ for 117.6.

§ 5. EFFICIENCY AND POWER OF BLADES. 93

This gives the following five final equations:

For No. 1, $\quad U = 2301 \dfrac{s - .0091}{(1-s)^2}.$

For No. 2, $\quad U = 2448 \dfrac{s - .0088}{(1-s)^2}.$

For No. 3, $\quad U = 2237 \dfrac{s - .0093}{(1-s)^2}.$

For No. 4, $\quad U = 2420 \dfrac{s - .0088}{(1-s)^2}.$

For No. 5, $\quad U = 2557 \dfrac{s - .0086}{(1-s)^2}.$

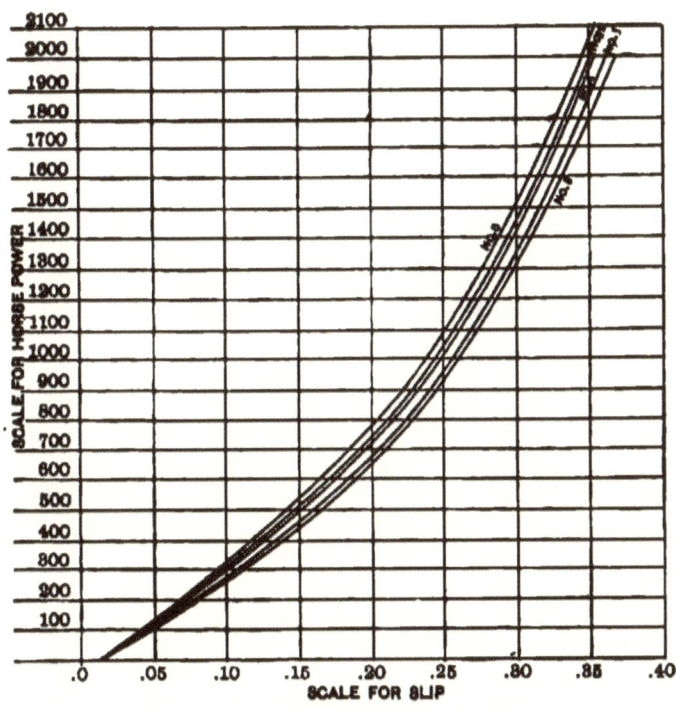

Fig. 33. — HORSE-POWER DELIVERED AT CONSTANT SPEED OF ADVANCE.

Figure 33 shows the curves of useful horse-power obtained from the above, plotted on slip as before.

Their shape is essentially the same as that of curves of useful horse-power of actual model propellers obtained by experiment.

When I come to discuss the question of propeller design I shall have something more to say upon the theory of the propeller acting alone. At present, however, I shall leave this subject and discuss the mutual reactions of ship and propeller when the latter is driving the former.

CHAPTER III.

MUTUAL REACTIONS BETWEEN PROPELLER AND SHIP.

§ 1. *Action of Propeller attached to Vessel.*

A little consideration will make it evident that a screw attached to a ship, and driving it through the water, is working under conditions decidedly different from a screw advancing through undisturbed water in the direction of the axis of its shaft. The resistance of the ship also must be affected by the working of the screw. I shall consider first the case of the screw.

Even if the screw of a ship is of correct and uniform pitch, its shaft is seldom exactly parallel to the centre line of the ship, and to the undisturbed surface of the water. Hence when the screw is propelling the ship in smooth water, it is not in general advancing parallel to its shaft. The result is that the slip angle varies from point to point of a revolution.

Screw seldom advances Parallel to itself.

The variation (supposing the water undisturbed) follows the following simple law.

Let ϕ_0 denote the mean slip angle of an element, and $2\phi_1$ the total amount of variation of the slip angle during a revolution. Let ψ denote the angle made by a fixed radius of the shaft with a fixed direction in space, and ϕ the actual slip angle at the point of the revolution determined by ψ.

Then $$\phi = \phi_0 + \phi_1 \sin \psi.$$

If ϕ_1 is not large in comparison with ϕ_0, the slip will vary with sufficient approximation directly as the slip angle, or $s = s_0 + s_1 \sin \psi$, where s, s_0, and s_1 apply not only to a single element, but to the whole propeller.

So much for deviation of the shaft from the true level and fore-and-aft line.

Disturbances of Water of Screw Race.

Now the water in which the screw works is disturbed by the action upon it of the ship. The disturbance is complex and has a powerful influence upon the screw's action. This complex disturbance may be regarded as made up of four comparatively simple components.

Non-parallelism of Stream Lines.

a. The water closing in around the stern tends to flow parallel, or nearly so, to the surface of the hull, and hence to flow at an angle to the fore-and-aft vertical plane. This manner of flow will have but little effect upon the propeller of a single-screw vessel, since in the centre line the water tends to flow parallel to the fore-and-aft plane. With twin-screw ships the result of the water flowing parallel to the hull is to produce a virtual deviation of the shaft.

Stream Line Wake.

b. Owing to stream line action the velocity of the water in the neighbourhood of the stern is less than the normal speed.

Figure 2 illustrates this fact, and shows also that the velocity will not be the same throughout the screw's disc.

For a ship advancing through still water this stream line action tends to produce a following current or wake, which is not uniform over the disc of the screw. I shall call it the "stream line wake."

Frictional Wake.

c. Owing to the "frictional wake," or the following current set in motion by the friction of the wetted skin, the uniformity of the race or column of water acted on by the screw is still further disturbed. The "frictional wake" is not a uniform current, but is strongest close to the hull, and near the surface, diminishing outwards and downwards.

Wave Motion Wake.

d. The presence of waves implies motion of the water, and hence disturbance of the race.

The disturbance due to a wave is the greater the nearer the surface, and varies in direction according as the propeller falls beneath a crest or a hollow.

A crest implies forward motion of the water; a hollow,

sternward motion. This disturbance I shall call the "wave motion wake."

Summing up, it is evident—

1. That the slip of the blade of a propeller driving a ship necessarily varies from point to point of a revolution. *Summing up.*

2. That the apparent slip or (speed of screw − speed of ship) ÷ (speed of screw) can be the same as the real slip only by chance.

In practice, the real slip is always greater than the apparent slip. The frictional and stream line wakes always tend to diminish the apparent slip, and even if the wave motion wake is opposed to the two above, it is not sufficiently strong to neutralise them.

All our theoretical conclusions concerning the screw have been based upon the assumption that it advanced parallel to the axis of the shaft into undisturbed water. We see that the actual working conditions are very different. If, then, we are to make any practical use of previous results, it is necessary to bridge the gap between theory and practice by a reasonable working hypothesis.

While at a given point of a revolution the slip varies from point to point over a blade, the variation, with the amounts of slip met with in practice, must be small, and it seems entirely allowable to assume a mean slip applicable to the entire blade at a given point of its revolution. *Expression for Slip at Any Point.*

Now for the screw working in the disturbed wake, the slip of a blade must be at a maximum at some point of the revolution, and at a minimum at some other point. Also from the nature of the motions these points must be nearly opposite one another, *i.e.* about 180° apart. Furthermore, at points half-way between the point of maximum slip and the point of minimum slip, the slip cannot be far from the mean between the maximum and minimum.

Let ψ denote the angular distance of a blade from the position corresponding to the mean slip. Let s denote the slip corresponding to ψ, s_0 the mean slip, and $2s_1$ the difference between the maximum and minimum slip. Then it follows from the above that with fair approximation

$$s = s_0 + s_1 \sin \psi.$$

Gross and Net Work in Turbulent Wake.

Referring to page 80, it is evident that we may denote N', the gross work delivered to a blade in one revolution at slip s, by $K(asX_e + fZ_e)$, and the useful work U' by $K(1-s)(asX_e - fY_e)$.

While turning through a small angle $d\psi$ at slip s, the elementary useful work $= \dfrac{N' d\psi}{2\pi}$, or

$$dN' = \frac{K}{2\pi}(asX_e + fZ_e)d\psi,$$

$$dU' = \frac{K}{2\pi}(1-s)(asX_e - fY_e)d\psi.$$

Substituting in the above the value of s in constants and terms involving ψ, namely,

$$s = s_0 + s_1 \sin \psi,$$

we have

$$dN' = \frac{K}{2\pi}(as_0 X_e + fZ_e + as_1 X_e \sin \psi)d\psi,$$

$$dU' = \frac{K}{2\pi}\left\{ \begin{array}{l}(1-s_0)(as_0 X_e - fY_e) + (1-s_0)as_1 X_e \sin \psi \\ -s_1 \sin \psi (as_0 X_e - fY_e) - as_1^2 \sin^2 \psi X_e \end{array} \right\} d\psi.$$

Integrating for one revolution between the limits 0 and 2π, we have

$$N' = K(as_0 X_e + fZ_e),$$

$$U' = K\left[(1-s_0)(as_0 X_e - fY_e) - \frac{aX_e}{2}s_1^2\right].$$

Hence the gross work absorbed by the blade is that corresponding to the uniform slip s_0, while the useful work is diminished because of the negative term $\dfrac{aX_e}{2}s_1^2$.

ACTION OF PROPELLER.

The diminution is small, however, unless s_1 is large compared with s_0.

This question of the effect of non-uniformity of wake, or turbulence of wake, upon the action of the propeller is a most important one. For that reason a second method of dealing with it may well be introduced.

Suppose that of the total work W sent to the propeller, an amount w_1 is expended at the slip s_1, with an efficiency e_1, an amount w_2 at slip s_2, with efficiency e_2, and so on.

Efficiency by Second Method in Turbulent Wake.

Then the useful work

$$U = w_1e_1 + w_2e_2 + w_3e_3.$$

The efficiency of the whole transaction

$$= \frac{U}{W} = \frac{w_1e_1 + w_2e_2 + w_3e_3 + }{w_1 + w_2 + w_3 + }.$$

Also the mean slip s_0

$$= \frac{w_1s_1 + w_2s_2 + w_3s_3 + }{w_1 + w_2 + w_3 + }.$$

Suppose next that the curve of efficiency of the propeller plotted on slip is a straight line, or that $e_1 = a + cs_1$, $e_2 = a + cs_2$, and so on. Then

$$\frac{U}{W} = E = \frac{w_1(a+cs_1) + w_2(a+cs_2) + w_3(a+cs_3) + }{w_1 + w_2 + w_3 + }$$

$$= \frac{a(w_1 + w_2 + w_3 + \cdots)}{w_1 + w_2 + w_3} + c\frac{w_1s_1 + w_2s_2 + w_3s_3 + \cdots}{w_1 + w_2 + w_3}$$

$$= a + cs_0.$$

That is to say, the final efficiency of transfer corresponds to the mean slip if the efficiency curve is a straight line.

Now, unless the slip variations are large, the efficiency curve within the limits of variation is practically a straight line. In such a case the efficiency is that corresponding to the mean slip.

Conclusions.

The above theoretical work appears to justify the following conclusions:

1. The variable wake currents around a propeller are equivalent to a single uniform wake.

Froude's Experimental Conclusions.

2. The efficiency of a propeller is not appreciably affected by the turbulence or variation of the wake unless the variation is excessive.

These conclusions fully agree with the results of model experiments made by the Froudes upon small propellers working behind models of ships.

R. E. Froude has stated that in such experiments the turning moment and thrust correspond to the mean slip, and that the experiments appeared to indicate, if anything, a very slight gain of efficiency due to turbulence of wake.

While moderate variation in the wake does not affect the turning moment, thrust, and efficiency of the propeller as a whole, it should be remembered that it causes great variations in the thrust of a given blade during a complete revolution.

Virtual and Actual Deflection of Shafts.

In practice, the wake is always greatest near the surface. For a twin-screw ship it is also greater in the part of the disc close to the ship than in the part farthest removed. Hence, in a twin-screw ship with propellers turning outward (*i.e.* with the blades moving away from the ship when in their highest position), the wake is equivalent to a virtual deflection of the shaft (starting from the engine) outward and upward. If the propellers turn inward, the virtual deflection is inward and downward. Care should be taken, in designing, not to give the shaft an actual deflection of such a nature as to add to the virtual deflection due to the wake.

On the contrary, the virtual deflection should, in the case of twin-screw vessels, be more or less neutralised by a slight actual counteracting deflection.

§ 2. *The Wake.*

Having concluded that the propeller behind a ship is practically working in a uniform following current or wake, it is necessary to inquire into the effect of this wake upon some of our previous results and formulæ.

It is convenient to express the speed of the wake as a fraction of the speed of the ship. Let us denote it by wV, where w is called the "wake fraction," or "wake factor." Retaining the symbol s to denote true slip, it is convenient to denote apparent slip by the symbol s'. The relation between s and s' must involve w. *Wake Factor. Real and Apparent Slip.*

Now, speed of screw in knots $= \dfrac{pR \times 60}{6080}$,

speed of ship in knots $= V$,

speed of wake in knots $= wV$.

$$\text{Apparent slip} = s' = \dfrac{\dfrac{pR \times 6}{608} - V}{\dfrac{pR \times 6}{608}} = 1 - \dfrac{608 V}{6 pR}.$$

$$\text{True slip} = s = \dfrac{\dfrac{pR \times 6}{608} - (1-w)V}{\dfrac{pR \times 6}{608}} = 1 - \dfrac{608 V}{6 pR}(1-w).$$

Whence, $\qquad (1-s') = \dfrac{(1-s)}{1-w}$,

or, $\qquad s = w - s'(1-w), \quad s' = \dfrac{s-w}{1-w}, \quad w = \dfrac{s-s'}{1-s'}$.

What is the effect of the wake upon the gross and useful work?

The thrust and the work delivered to a screw depend solely upon the revolutions and the speed of advance; *i.e.* upon the true slip. The useful work, however, depends upon the velocity with which the thrust overcomes the *Thrust and Work considering Wake.*

resistance opposing it; and hence, upon the speed of the ship. Referring to page 71, it is evident that, in deducing the useful work done by a propeller attached to a ship, we must multiply the thrust by the speed of the ship, that is, by $pR(1-s')$, instead of by the speed of advance of the propeller, which is denoted by $pR(1-s)$.

The only change in the final formulæ is in the expression for useful work, which becomes

$$U = 3n\left(\frac{pR}{1000}\right)^3 d^2 (1-s')(asX_c - fY_c),$$

while for the gross work N we still have

$$N = 3n\left(\frac{pR}{1000}\right)^3 d^2 (asX_c + fZ_c).$$

Efficiency Considering Wake. Consider next the effect of the wake upon efficiency. The expression for efficiency from page 74 is

$$e = (1-s)\frac{cs - Y_c}{cs + Z_c};$$

a consideration of the deduction of this expression will show that when the effect of the wake is included, it becomes

$$e' = (1-s')\frac{cs - Y_c}{cs + Z_c};$$

but
$$1 - s' = \frac{1-s}{1-w},$$

so we can also write the expression for efficiency when wake is considered,

$$e' = \frac{1-s}{1-w} \times \frac{cs - Y_c}{c_s - Z_c},$$

where s is as usual the true slip.

Now e' is an apparent, not a real, efficiency. It is the ratio between the work delivered to the ship by the propeller and the work delivered to the propeller by the shaft, but a certain amount of the work delivered to the ship by the propeller is obtained from the water of the wake, instead of from the shaft. A simple ideal case may help to make clear this somewhat difficult point. Imagine a propeller working in a canal, and driving a carriage on rails above the canal. Suppose the carriage so fitted with adjustable brakes that its resistance, and hence the thrust required to drive it, is constant for a wide range of speed, say from 5 to 20 knots. Suppose now such a turning moment applied to the propeller that it drives the carriage at 5 knots, the water in the canal having no motion.

Keeping the same turning moment, suppose the water of the canal to be flowing at the rate of 5 knots. The thrust of the propeller will be the same as before, and the carriage will now move at 10 knots. The power applied to the propeller will be the same, and the portion of this power utilised for driving the carriage will be unchanged; but an additional power equal to the latter is delivered to the propeller by the water in the canal.

Under these conditions the "negative slip" may be large, and the "apparent efficiency" much above unity.

This brings me to the much-vexed question of "negative slip" in the case of actual propellers. *Negative Slip.*

It is absurd to discuss the possibility of real negative slip.

A propeller cannot exert thrust without slip.

But suppose we had a propeller, which, with a speed $\left(\dfrac{6pR}{608}\right)$ of 10 knots and a slip of 10% (making a speed of advance of 9 knots), exerts in still water sufficient thrust to overcome the resistance of a certain ship at 11 knots.

Suppose the ship has a wake of 2 knots, and that the propeller is put in the ship and driven by power sufficient to give it the same number of revolutions as before. The speed of the propeller, as deduced from the pitch multiplied by revolutions, will be 10 knots, the thrust will be that necessary to drive the ship at 11 knots, and she will undoubtedly show that speed, involving an apparent negative slip of 10%. The possibility of negative slip depends, then, entirely upon the possibility of a wake of a speed greater than the actual speed of slip. The existence of such an amount of wake in many cases of full-sterned ships, where a large stream line wake is added to the frictional wake, is incontestable, and there appears to be no reason why reports of apparent negative slip should be received with discredit and contumely.

Apparent negative slip is, however, very objectionable, as it is always a sign of low actual slip, and hence of low propeller efficiency.

§ 3. *Thrust Deduction.*

Cause of Thrust Deduction. When a propeller works with slip, the faces of the blades exert pressure upon the water, and hence increase the pressure aft of the propeller.

Also the backs of the blades exert suction upon the water, and hence decrease its pressure forward of the propeller. If the propeller is attached to a ship, this decrease of pressure extends to the water in contact with the hull, decreases the pressure upon the after-body, and hence increases the resistance.

This increase of resistance was called by the elder Froude the augment of resistance, but has been more happily termed the thrust deduction by his son.

Influences causing Variation of Thrust Deduction. It does not appear possible that in any case the thrust deduction can exceed that part of the thrust of a screw due to the suction of the backs of the blades, and it must in most cases be much less.

The amount of the thrust deduction for a given ship must vary with the position of the screw relative to the hull.

The thrust deduction may be expected to change but little with change in size, proportions, or slip of the screw, provided its thrust remains unchanged.

Our knowledge upon the question of thrust deduction is due almost entirely to the researches of the Froudes. It is, however, exceedingly limited. I shall give subsequently some of the results published by the Froudes, which are found of use in considering practical cases.

CHAPTER IV.

ANALYSIS OF TRIALS AND AVERAGE RESULTS.

§ 1. *Value of Trials.*

Indicated and Effective Horse-Power Differ.
The power indicated by the engines of a ship and the power required to overcome the tow rope resistance are very different. The latter absorbs in practice only from 40 to 60 per cent of the former, the remainder being accounted for in various ways. Evidently the analysis or separation of the indicated horse-power into its several components is very desirable, and may in fact be said to be necessary if the results of trials of actual ships are to be fully utilised in preparing designs for new ships.

Mercantile Value of Trials.
A careful analysis of trial results of a new ship is of value also from another point of view. It enables the owner to decide whether or not the performance of his ship is what it should be, and may point out defects that can easily be made good. In the long run the results of such defects would, in the case of a merchant vessel, show on the owner's books, but the books would seldom indicate the remedy.

Graphic Method used in Analysis.
Before taking up the question of trial analysis I shall explain a graphic method which I shall have occasion to use. In dealing with trial analysis, where the quantities which we must handle are at best approximate, graphic or semi-graphic methods present distinct advantages over arithmetical or algebraical processes.

§ 1. VALUE OF TRIALS. 107

Now if the positions of these lines had to be obtained by experiment or observation, we should find that they would not all pass through a single point. We might know that if our observations or experiments were exact, the lines would have a focus. But it is not possible for fallible man to entirely avoid errors of observation. Refined methods and apparatus combined with skill may, and often do, so bring it

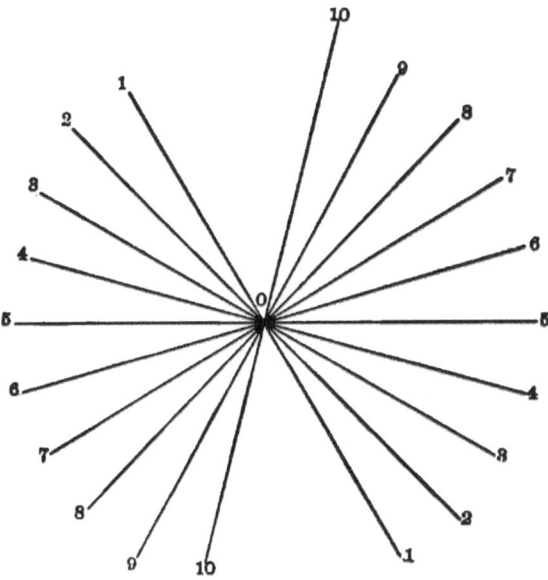

Fig. 34. — LINES RADIATING FROM A FOCUS.

about that errors of observation may be neglected for any practical purpose. They cannot, however, be eliminated entirely.

In Figure 35 the lines of Figure 34 are shown with small errors in position assigned at random. The focal point of Figure 34 is indicated in Figure 35 by a small circle. While few of the lines pass directly through this focal point, it is evident that any one who knew that all the lines should pass through a focus, could not go far wrong in "spotting" it upon

Figure 35. If the errors of the lines of the diagram were not abnormal (a fact which the diagram itself would show), the focal point so spotted would be quite sufficiently close to the actual focus for all practical purposes.

The reader will be enabled to form his own conclusions as to the accuracy of this graphic method from examples of its application which I shall introduce presently.

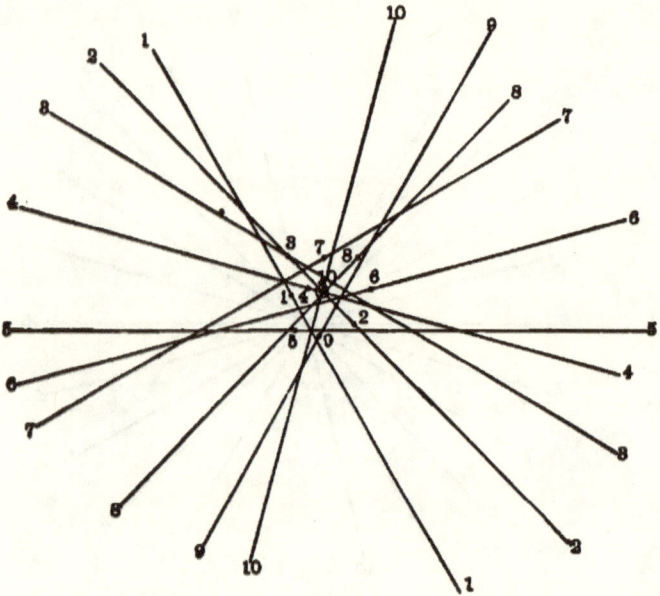

Fig. 35. — LINES RADIATING FROM APPROXIMATE FOCUS.

§ 2. *Components or Absorbents of the Indicated Horse-Power.*

Results of Progressive Speed Trials.
A discussion of the manner of conducting a "progressive trial" of a steamship is beyond the scope of this work. Suffice it to say that in such a trial the indicated horse-power and the revolutions of the engines are determined for various speeds ranging from four or five knots up to the highest speed of which the ship is capable. It is best in dealing with the results of trials to lay down the revolutions as abscissæ, set

§ 2.   COMPONENTS OR ABSORBENTS.   109

up ordinates representing on suitable scales the corresponding speed in knots and indicated horse-power, and draw through the extremities of the ordinates fair curves of speed and indicated horse-power. For twin-screw vessels the revolutions are taken as the average of the two screws. For triple-screw vessels, where the middle screw, even if of the same

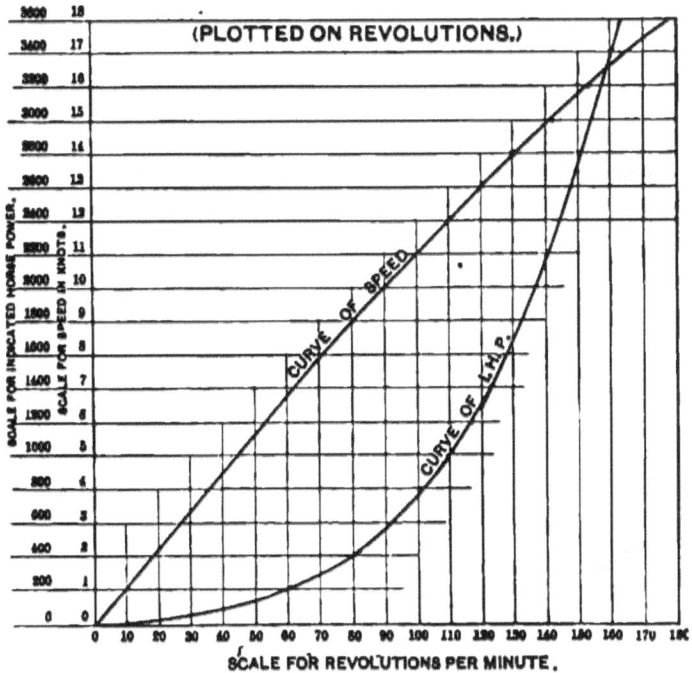

Fig. 36. — CURVES OF SPEED AND I. H. P. OF U. S. S. YORKTOWN.

size as the other two, is working under different conditions, it will be necessary to draw separate curves for the middle screw.

Figure 36 shows curves of speed and indicated horse-power, plotted as indicated above from the trial results of a twin-screw vessel, the United States gunboat *Yorktown*.

Evidently from Figure 36 we can determine the revolutions and horse-power corresponding to any speed.

Let us trace the processes by which a greater or less proportion of the horse-power indicated in the cylinders of the engines is used to overcome the resistance of the ship.

Reduced Mean Effective Pressure. Knowing the cylinder dimensions of the engine, we can reduce the mean effective pressures obtained in the several cylinders to an equivalent mean effective pressure upon the low-pressure piston only. Thus in the case of the *Yorktown* the high-pressure cylinder of each engine is 22 inches in diameter, the intermediate 31 inches, and the low-pressure 50 inches.

Then to reduce an effective pressure shown by an indicator diagram in the high-pressure cylinder to an equivalent effective pressure in the low-pressure cylinder, we must multiply it by the factor $(\frac{22}{50})^2 = .1936$. Similarly, for the intermediate cylinder the factor of reduction is $(\frac{31}{50})^2 = .3844$.

Expression for Indicated Horse-Power. Let p_m denote the total equivalent mean effective pressure, reduced to the low-pressure cylinder.

Let R denote the revolutions.

Let I denote the indicated horse-power.

Let C denote the horse-power constant for the low-pressure cylinder.

Then
$$I = C \cdot p_m \cdot R.$$

C depends upon the cylinder dimensions. If A denote the effective cylinder area in square inches, and s the stroke in feet,
$$C = \frac{2As}{33000}$$

From the above relation between I, p_m, and R, it is evident that if either R or p_m is equal to zero, I will become equal to zero.

Now at the origin, $R = 0$, and hence $I = 0$. It does not follow that $p_m = 0$. On the contrary, p_m must have a value.

Initial Friction. Owing to the tightness of glands and bearings, pistons and valves, a certain amount of effective pressure is necessary to start the engine against its own internal resistance — to overcome the "initial friction" or "dead friction." This initial friction or internal resistance of the engine is overcome at all rates of revolution. While it probably varies slightly with the power and revolutions, we have no reason to believe the variation great.

It is then safe to suppose this resistance constant, and hence the power absorbed by it varying as the revolutions.

If p_0 denote the mean effective pressure necessary to overcome the initial friction, and I_f the corresponding horse-power absorbed at R revolutions, we have $I_f = C p_0 R$.

At the origin p_m and p_0 coincide, and hence the curve of I coincides with the curve of I_f. This is because with the exception of I_f, all the resistances which absorb I diminish indefinitely with the speed, and become zero when the speed and revolutions become zero.

If we know p_0, we can calculate I_f for any value of R, and hence for any speed.

To determine p_0, we proceed as follows: It is supposed that in properly conducted speed trials, "spots" on the power curve are determined at close intervals down to the lowest rate of revolutions which can be steadily maintained by the engines. These spots being set off in both the first and third quadrants, the curve of I. H. P. is plotted through the origin, as shown in Figure 37. From this curve we plot a derived curve of (I. H. P.) ÷ (Revolutions), as shown in the figure. Spots for this curve are obtained by dividing the I. H. P.'s, corresponding to various rates of revolutions by the revolutions. The spots of this curve should be laid off in both the first and second quadrants, as the curve is horizontal immediately above the origin and symmetrical on either side of the vertical axis.

Of course we cannot calculate the ordinate for the spot where the vertical axis is cut, since there $I = 0$, $R = 0$. We can, however, obtain spots so close to it on either side that by sweeping a fair curve through — horizontal where it cuts the axis

Determination of Initial Friction.

—the value of $\frac{I}{R}$ for $R=0$ is very accurately determined. From the curve of $I \div R$ a curve of p_m is readily drawn, since

$$I = Cp_m R,$$

whence
$$p_m = \frac{1}{C}\frac{I}{R}.$$

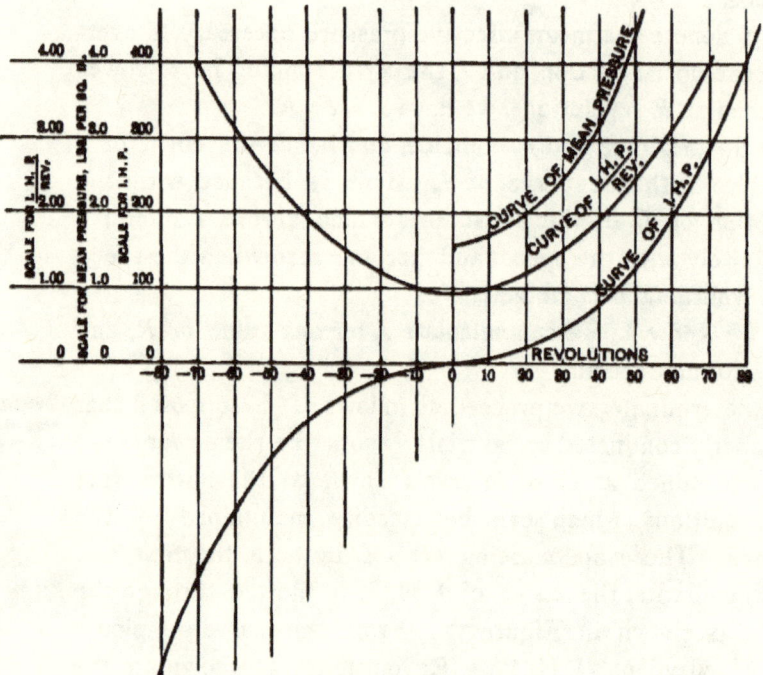

Fig. 37.—CURVES OF $\begin{cases} \text{I. H. P.} \\ \text{MEAN PRESSURE.} \\ \text{I. H. P.} \\ \text{REVOLUTIONS.} \end{cases}$ PLOTTED ON REVOLUTIONS.

For the *Yorktown* the low-pressure cylinder area is 1963.5 inches, and the stroke 2.5 feet; whence

$$C = \frac{2 \times 1963.5 \times 2.5}{33000} = .2975,$$

$$\frac{1}{C} = 3.361.$$

From the curve, the value of $\frac{I}{R}$ for $R=0$ is .95.

Then (remembering that there are two low-pressure cylinders),

$$p_o = \frac{1}{2} \times .95 \times 3.361 = 1.6.$$

Figure 37 shows a curve of p_m, including p_0. If it is more convenient, this curve may be drawn at once instead of the curve of $I \div R$.

But while apparently a more round-about process, it is in my opinion, simpler and better to determine p_0 from the auxiliary curve of $I \div R$.

We thus determine directly the quantity Cp used to calculate I_f by the formula $I_f = Cp_0 R$.

Having calculated the values of I_f, we denote $I - I_f$ by I_ω and call it the net horse-power.

It may be remarked here that the value of the initial friction, in the case of *Yorktown*, as determined above, is very low for new engines. At full speed the initial friction absorbs only about $4\frac{1}{2}$ per cent of the total indicated horse-power. In new triple-expansion engines of the present day, I_f usually absorbs from 6 to 10 per cent of the total power. An amount of initial friction greater than 10 per cent can always be traced to avoidable causes — such as bad fitting or poor alignment. In cases where feed, air, circulating or bilge pumps are driven by the main engines, the power required to work them should be determined and classed with the initial friction. At the present day, in few large ships do we find any pumps except the air-pump worked as above. I shall give later an estimate by a competent authority of the average amounts of power absorbed by the various pumps — when worked off the main engines.

Besides the loss through "initial friction," there is a second frictional absorbent of power. This is the "load Load Friction.

friction," being the friction of the load upon the bearings and the thrust block. It necessarily varies with the coefficient of friction of the rubbing surfaces, and hence depends somewhat upon the lubrication. There are scarcely any reliable experimental results from which the load friction of marine engines under various conditions can be certainly predicated. It appears to be nearly always about 7 per cent of the net horse-power, $I_n(=I-I_i)$ seldom falling below 5 per cent. All things considered, an allowance of 7 per cent of the net horse-power for load friction appears to be reasonable, and I shall adopt it in future. Mr. Isherwood, the distinguished American engineer, considers $7\frac{1}{2}$ per cent the proper allowance. Mr. Blechynden, of Barrow, gives $6\frac{1}{5}$ to $7\frac{1}{2}$ per cent for triple-expansion engines of the present day. The horse-power absorbed in load friction I shall denote by I_t.

Power and Efficiency of Propeller. We have now traced the power to the propeller. Denoting the power delivered to the propeller by P, we have

$$P = I - I_t - I_l.$$

Of P a certain amount is wasted by the slip of the propeller, and a certain amount is wasted by the friction (including head resistance) of the propeller blades.

Denoting the sum of these two losses by P_s, we conclude that an amount of power derived from P equal to $P - P_s$ is delivered to the thrust block — the true efficiency of transfer being $\dfrac{P - P_s}{P}$.

Wake Gain. We have seen also that a certain amount of power is delivered to the propeller by the wake and transferred to the thrust block, so that the actual amount of power absorbed by thrust is greater than $P - P_r$.

Denote the actual power absorbed by thrust by T; we have the apparent efficiency of the propeller equal to $\dfrac{T}{P}$

Now we have seen (page 102) that if e' denote the apparent efficiency of a propeller, with wake factor $= w$ and e denote the true efficiency,

$$e' = \frac{e}{1-w}$$

Now
$$e' = \frac{T}{P},$$

$$e = \frac{P-P_i}{P};$$

whence
$$T = \frac{P-P_i}{1-w}$$

I shall denote the power absorbed by the propeller from the wake by P_w. Then we have $T = P - P_i + P_w$.

Of T, the power absorbed by the thrust, a certain amount is used to overcome the thrust deduction or suction, exerted by the propeller upon the rear of the hull. Let us denote the thrust deduction by T_i. Then if t denote the thrust deduction factor, we have $T_i = tT$. **Thrust Deduction**

The remaining portion of T is employed to overcome the natural or tow rope resistance of the ship. This portion of the power alone is used to produce the effect for which the power is needed. Hence it is well called the "effective horse-power," and denoted by E. **Effective Horse-Power.**

As we have distinguished two components of the resistance (R), namely, R_s the skin resistance, and R_w the wave resistance, it is necessary to divide the effective horse-power into two corresponding portions. **Skin and Wave Resistance Power.**

We will denote them by E_s the skin resistance power, and E_w the wave resistance power.

Recapitulating in symbols, we have,

$$I = I_f + I_i + P,$$
$$P + P_w = P_i + T,$$
$$T = T_i + E,$$
$$E = E_s + E_w.$$

With a satisfactory method of trial analysis, all of the quantities denoted above can be determined with sufficient aproximation from known data concerning the ship and engines, and the results of careful progressive trials.

§ 3. *Yorktown Trial Analysis.*

We have seen already how to determine I_f in a given case, and also that with sufficient approximation,

$$I_t = .07 (I - I_f).$$

I can best make clear the methods pursued in determining the other components of the indicated horse-power, by analysing in detail the trial results of the *Yorktown* — shown in Figure 36.

Data of Yorktown and Formulæ needed.

The trials of the *Yorktown* were carefully conducted by a board of naval officers, and the accuracy of the results can be relied on. The trials were made in 1889, shortly after the commissioning of the vessel. Very complete particulars of the ship and machinery and of the trials have been published in the Journal of the American Society of Naval Engineers, and official reports of the U.S. Navy Department. Those needed for the purpose in hand are given below:

Length between perpendiculars, 226'.

Length on trial water line, 230'.

Mean immersed length, 226'.

The vessel has a slight partially immersed overhang, but it did not appear to be large enough to warrant making the mean immersed length greater than the length between perpendiculars.

§ 3. YORKTOWN TRIAL ANALYSIS.

Beam extreme on trial water line, 36'.
Draught on trial forward, 12' 6"½.
Draught on trial aft, 15' 0"¾.
Draught on trial mean, 13' 0".82.
(This is skin draught.)
Displacement on trial, 1680 tons.

The displacement and trim on trial were practically identical with the displacement and trim at designed load water line.

Wetted surface (S) on trial, 10,840 square feet.
Area of midship section (M) on trial, 432 square feet.

The propelling machinery is of the twin-screw, horizontal, triple-expansion, three-cylinder type, for boiler pressure of 160 pounds above the atmosphere.

The cylinders are 22, 31, and 50 inches in diameter by 30 inches stroke.

The propellers are three-bladed, the blades being of the oval or modified Griffiths shape.

The pitch (uniform), 12' 6".
Diameter of propeller, 10' 6".
Diameter of boss, 2' 6".

The blade area of each propeller is 25.4 square feet.
The mean width ratio = .201.
From the shapes of the blades the characteristics are found to be as follows:

$$X_c = .0954$$
$$Y_c = .1439$$
$$Z_c = .4469$$

From Table V. we obtain .0094 as the appropriate skin resistance coefficient for the *Yorktown*.

Then $R_r = .0094 \times 10840\, V^{1.88}$.

R_r is in pounds. Now a resistance of one pound at a speed of one knot absorbs .0030707 horse-power.

Then $E_r = .0030707\, V \times R_r$

$\qquad = .0094 \times 10840\, V^{1.88} \times .0030707\, V$

$\qquad = .3129\, V^{2.88}$.

Denoting the wave resistance R_w by bV^4, we have

$$E_w = bV^4 \times .0030707\, V = .0030707\, bV^5.$$

We have seen that the expression for the power P absorbed by a screw is

$$P = 3n\left(\frac{pR}{1000}\right)^3 d^2 (asX_c - fZ_c).$$

In the case of the *Yorktown* with three-bladed twin screws,

$$n = 6, \quad p = 12.5, \quad d = 10.5, \quad X_c = .0954, \quad Z_c = .4469.$$

Taking the standard value .045 for f, we obtain, upon simplifying the expression for P,

$$P = 369.58\left(\frac{R}{100}\right)^3 (as + .211).$$

The first step of the analysis is the determination of horse-power wasted per revolution by the dead friction. The method to be followed has already been explained, Figure 37, referring to the *Yorktown*, and giving us the result $I_f = .95\, R$.

NOTE.— The diagram and figures used in trial analysis are drawn in practice upon a much larger scale than those in the accompanying plates. The latter have been reduced from the large-sized figures, which must be used if accurate results are desired.

Table.

Turning now to Table X., it is seen that on the right, under the heading "Remarks and Explanations," are entered the necessary data and formulæ, together with results of analysis.

§ 3. YORKTOWN TRIAL ANALYSIS.

In line No. 1 we enter as headings for the columns the speeds. Lines 2 and 3 contain the corresponding values of the revolutions R, and the indicated horse-power I. They are obtained from the curves of Figure 36. While it would have been possible to extend the table to speeds below five knots, it would be unnecessary labour — the initial friction having been already determined.

Knowing that $I_f = .95 R$, we fill up line 4 by entering the products of the values of R in line 2 by .95. Line 5 gives the values of $I - I_f$, which are obtained by subtracting the quantities in line 4 from those in line 3. With the standard loss of 7 per cent for load friction, the power P delivered to the propeller is

$$(1.00 - .07)(I - I_f) = .93 \, (I - I_f).$$

The values of P so obtained are entered in line 6. We have seen also that

Wake Factor and Thrust Coefficient.

$$P = 369.58 \left(\frac{R}{100}\right)^3 (as + .1055).$$

Line 7 gives the values of $\left(\frac{R}{100}\right)^3$ obtained from line 2. Then line 8 is obtained by the use of the formula

$$as + .211 = P \div 369.58 \left(\frac{R}{100}\right)^3.$$

Deducting .211 from the quantities in line 8, we have in line 9 the values of as. Here s is the true slip which we do not know, because we do not know the wake factor. Also we do not know the value of a. In line 10 is entered the speed of the screw in knots $= \frac{60 p R}{6080} = .1234 R$ for the $12'.5$ pitch of the *Yorktown* propellers.

In line 11 is entered the apparent slip obtained by the formula

$$s' = \frac{.1234 R - V}{.1234 R}.$$

The apparent slip s' is also the true slip s if the wake factor w is equal to zero.

For a wake of .10 we must add to the apparent slip at speed V

$$\frac{.10\ V}{.1234\ R}.$$

The quantities to be thus added are entered in line 12, and adding 11 and 12 we obtain in line 13 the true slip for a wake of .10.

Again adding lines 12 and 13, we obtain in line 14 the values of the true slip on the supposition of .20 wake. Now in line 9 we had the values of as, while lines 11, 13, and 14 give the values of s on the three suppositions of (1) zero wake, (2) .10 wake, (3) .20 wake; whence we enter in lines 15, 16, and 17 the corresponding values of a.

Thus we obtain for each speed three spots on a curve of values of a, plotted upon values of wake factor.

Since a and w are constant, or nearly so, the curves for the various speeds must either all coincide or else pass through a common focal point, corresponding to the constant values of a and w.

Figure 38 shows the curves in question. They do not tend to coincide, but clearly tend to pass through a common focus. Naturally we do not find a perfect focus, and it is necessary to "spot" the proper focal point.

This is done as indicated in the figure where

$$a = 7.8, \quad w = .083.$$

Closeness of Approximation. Up to this point but two approximate assumptions have entered into our work; namely, that 7 per cent of the net horse-power was expended in friction of load, and that the frictional coefficient for the propeller was .0225. While these have been selected as suitable standard values, and are probably fair approximations, it is proper to consider what effect would be produced in the values of a and w, determined as above by possible difference between the actual values of f and load friction, and the standard values adopted.

By giving to f and to the allowance for load friction values different from the standard values, it is found that for an

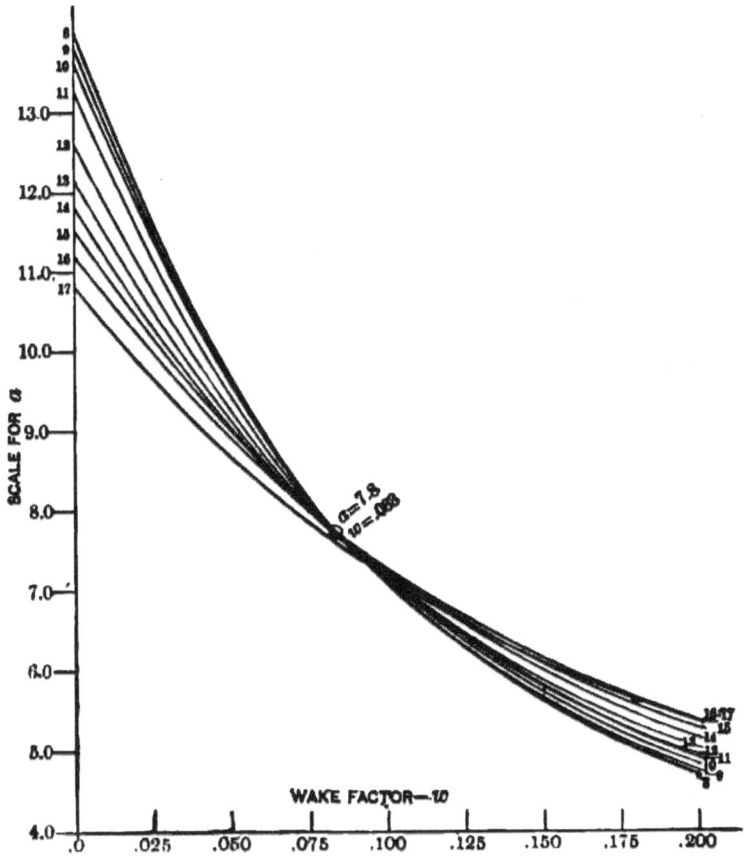

Fig. 38.— FOCAL DIAGRAM FOR a AND w.

increase in either quantity, the curves of Figure 38 are lowered bodily, and for a decrease are raised bodily. There is little or no change in the focal value of w.

While the focal value of a is changed, the change is small, for any probable variation of f or of the allowance for load friction from the standard values.

An error of 30 per cent in estimating the load friction causes an error of only about 2 per cent in the value of a. An error of 25 per cent in f, the coefficient for screw friction, results in an error of only about 3 per cent in the value of a.

All things considered, I think it safe to conclude that the method of Table X. and Figure 38 enables us, given accurate trial results, to determine the wake factor w very closely, and the thrust coefficient a with ample accuracy.

Knowing the wake, and hence the true and apparent slip, we are in a position to determine the true and apparent propeller efficiency, and the thrust horse-power T.

We have seen that the true efficiency e is given by the formula
$$e = (1-s)\frac{asX_e - fY_e}{asX_e + fZ_e}.$$

Rewriting,
$$e = (1-s)\frac{s - \dfrac{fY_e}{aX_e}}{s + \dfrac{fZ_e}{aX_e}}.$$

For the *Yorktown*
$$f = .045, \quad a = 7.8, \quad X_e = .0954, \quad Y_e = .1439, \quad Z_e = .4469.$$

Substituting and reducing,
$$e = (1-s)\frac{s - .009}{s + .027}.$$

Similarly, for the apparent efficiency e' we have
$$e' = (1-s')\frac{s - .009}{s + .027}.$$

Returning now to Table X., line 11ₐ, repeats line 11, the apparent slip. Line 18 gives the addition for the wake of .083.

Since line 12 gives the addition for a wake of .10, line 18 contains the quantities of line 12 multiplied by .83. Line

§ 3. YORKTOWN TRIAL ANALYSIS.

19, the sum of 11, and 18, gives the values of the true slip $=s$. Lines 20 and 21 contain respectively the values of $s-.009$ and $s+.027$. Then line 22 gives the apparent propeller efficiency, obtained from lines 11, 20, and 21 by the formula

$$e' = (1-s')\frac{s-.009}{s+.027}.$$

The thrust horse-power T, entered in line 23, is simply the product of the apparent efficiency of line 22 by the propeller power of line 6.

Now we have seen that a part of the thrust horse-power T is absorbed by T_t, the thrust deduction, the remainder being the effective horse-power E. Let t denote the thrust deduction factor;

Thrust Deduction.

then, $\quad T_t = T \times t;$

also, $\quad T = T_t + E = T \times t + E,$

or, $\quad T(1-t) = E.$

Now E is the sum of the skin resistance power E_s and the wave resistance power E_w. The former we know, it being equal to $.3129\ V^{2.83}$.

The latter is unknown, being denoted by $.0030707\ bV^6$, where b is an unknown semi-constant.

Then $(1-t)T = .3129\ V^{2.83} + .0030707\ bV^6$. Line 24 gives values of E_s or $.3129\ V^{2.83}$ calculated with the aid of Table XI. Line 25 is taken from Table XI., being values of $.0030707\ V^6$, denoted by y.

Let $\quad \dfrac{E_s}{y} = c,\ \dfrac{T}{y} = X.$

Then from the equation

$$(1-t)T = E_s + .0030707\ bV^6$$

$$X(1-t) = c + b.$$

Now t is practically constant, but we know that b varies more or less. If b and t were both constant, our task would be easy. We would have as many equations between them of the form $X(1-t) = c + b$ as there are speeds entered in the table, and their value could be determined graphically with great accuracy by plotting the focal diagram for the lines represented by the equations.

Unfortunately the case is not so simple. Remember that we are working with approximate quantities. Now at low speeds the wave-making resistance is so small, proportionally, that a small error in c (depending upon E_s), or in X (depending upon T), will render the equation $(1-t)X = c + b$ useless for a reliable determination of b. Small errors are necessarily made in such work as the present.

Again, at high speeds, E_w is substantial in comparison with E_s and T, but it is so large in comparison with T_i that the fluctuations, which we know take place in b, render the equation $(1-t)X = c + b$ unreliable for the determination of t.

So the determination of b and t must be made in some other way.

Average Equation for b and t.
We may, however, determine an average equation between b and t by determining average values of c and X for a suitable range of speeds. The upper speeds are preferable for this purpose. As noted in the table (under remarks, etc.), the average value of c for speeds from 11 to 17 knots inclusive, is .3563 and of X .6908. The corresponding equation is $.6908(1-t) = .3563 + b$.

If, then, we can in some way determine a fair average value of the fluctuating quantity b, we can from the above equation determine the value of the constant t, and hence the actual values at the various speeds of the fluctuating quantity b.

Let us then attempt the determination of b.

Returning to Table X., it is seen that line 28 contains the thrust efficiency $T \div I$.

§ 3. YORKTOWN TRIAL ANALYSIS.

If e denote the efficiency of propulsion, we have $e = \dfrac{E}{I}$.

Now $\quad E = (1-t)T.$

Then $\quad e = \dfrac{(1-t)T}{I} = (1-t) \times$ thrust efficiency.

Then $(1-t)$ being constant, the curve of true efficiency of propulsion is simply the curve of thrust efficiency on a smaller vertical scale. It follows that the ratio between the efficiencies of propulsion for any two speeds will be equal to the ratio between the thrust efficiencies for the same speeds.

Line 29 gives the ratio between the efficiency of propulsion at each tabular speed, and that at 16 knots, the highest even speed in knots actually attained by the *Yorktown*. The highest actual speed was about 16.7 knots, the quantities for 17 knots being obtained by extending the curves to that speed.

Returning to the expression for efficiency of propulsion, we have

$$e = \dfrac{E}{I} = \dfrac{E_s}{I} + \dfrac{E_w}{I}$$

$$= \dfrac{.3129\, V^{2.88}}{I} + \dfrac{.0030707\, V^5 b}{I}$$

$$= m + bn, \text{ say.}$$

Line 30 gives values of m, being obtained by dividing line 24 by line 3. Similarly, line 31 gives values of n, being obtained by dividing line 25 by line 3.

If $b=0$ the quantities of line 30 are the true efficiencies. Dividing them by the quantities of line 29, we get the values of e_{16} (or efficiency of propulsion at 16 knots) from each speed on the supposition that $b=0$. Line 33 gives the additions to be made to the values of e_{16} if $b=.5$. The quantities entered in it are obtained by multiplying line 31 by .5 (for $b=.5$) and dividing by the ratios in line 29.

Now from lines 32 and 34 we have a linear relation between e_{16} and b at each speed. Thus for 5 knots we have

$$e_{16} = .534 \text{ for } b = 0;$$
$$= .619 \text{ for } b = .5.$$

Then if we measure values of b. horizontally, and of e_{16} vertically, we can draw a straight line joining the points

$$e_{16} = .534, \ b = 0, \text{ and } e_{16} = .619, \ b = .5.$$

If our work were exact, and b were constant, we could draw a focal diagram with one line for each speed from which the values of e_{16} and b could be readily determined.

But it is an average value of a variable quantity which we wish to determine. Now the quantities in line 32 are slightly irregular because of the necessarily approximate nature of our work. The quantities in line 34 add to these irregularities, greater ones growing out of the fluctuations in the value of b.

The next step then is to spot the quantities of lines 32 and 34 above their proper speeds, and draw through the spots fair average curves.

These are shown in Figure 39.

Then from the fair curves new faired values of e_{16} for $b = 0$ and $b = .5$ are entered in lines 35 and 36.

Using the quantities in lines 35 and 36, the focal diagram of Figure 40 is drawn. This deals with values of e_{16} and b.

Second Approximation to b.

We have seen that for each speed the relation between e and b is expressed by

$$e = m + bn.$$

For each speed, then, we can plot a line, using the above equation. Figure 41 shows two such lines supposed to be for successive speeds of 14 and 15 knots.

Now if b is the same for the two speeds, while the efficiency is increasing from 14 to 15 knots, the true values of b and e

will evidently be greater than those corresponding to X, the point of intersection of the lines.

Similarly, if e is constant while b is increasing, the values of b and e corresponding to X are greater than the true

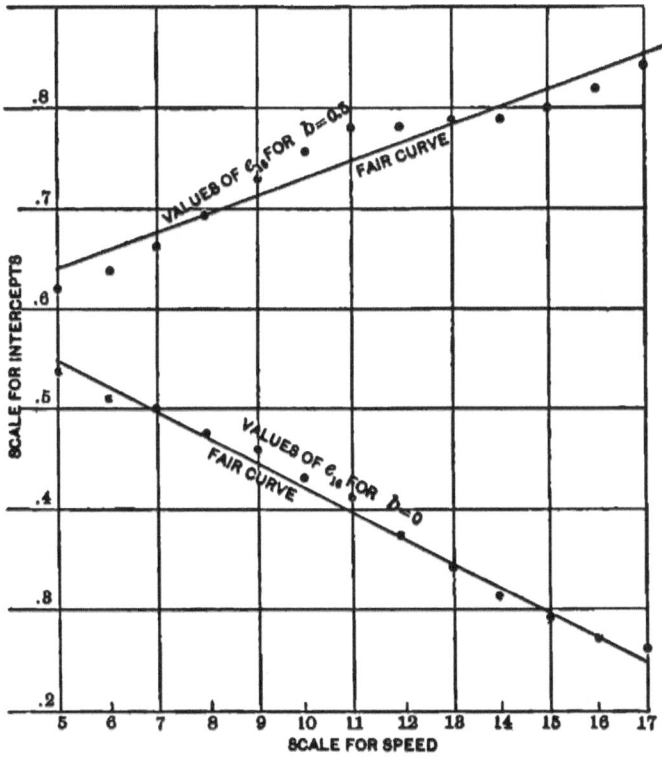

Fig. 39. — VALUES OF e_{16} FOR $b=0$ AND $b=5$ FAIRED.

values. If neither b nor e are constant, but both are increasing, the values corresponding to X are probably not far from the average of the values for the two speeds.

Now, in the case of the *Yorktown*, we know from line 29 that the efficiency increases always — rapidly at the lower speeds, slowly at the higher speeds.

Also, since at the highest speeds we have $\dfrac{V}{\sqrt{L}}$ greater than unity, it is certain that b is increasing at or near the top speed.

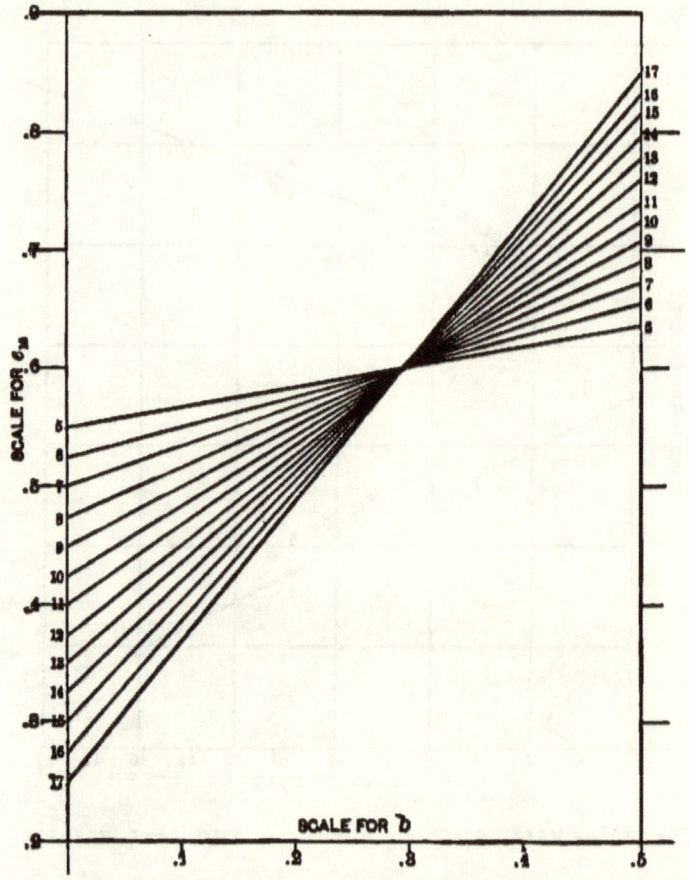

Fig. 40.—FOCAL DIAGRAM FOR b AND e_{16}.

So we may reasonably conclude that a diagram of lines

$$e = m + bn$$

will tend to show a focus at about a fair average value of b, especially for lines corresponding to the higher speeds.

§ 3.   YORKTOWN TRIAL ANALYSIS.   129

Figure 42 shows this diagram.

The ordinates for $b=0$ are simply the values of m in line 30.

For $b=.5$ we add to these ordinates .5 of the quantities in line 31.

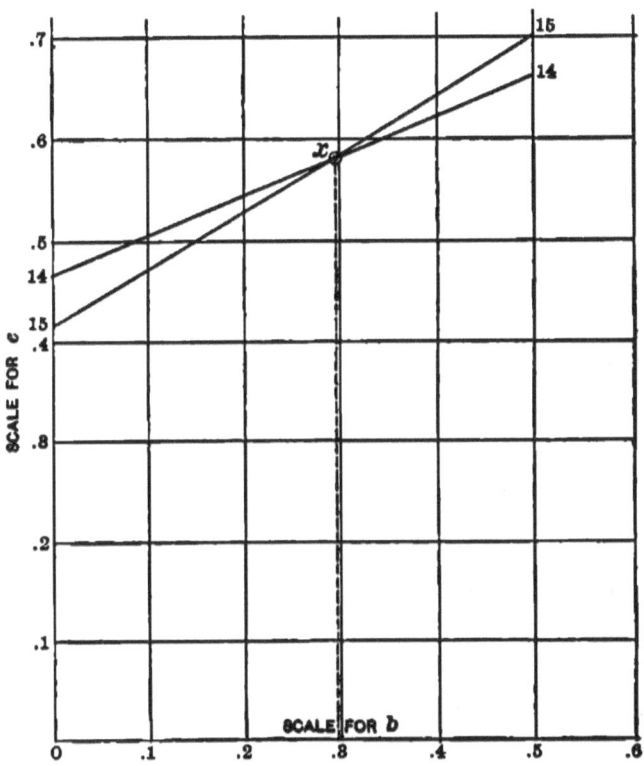

Fig. 41.— TWO LINES OF FOCAL DIAGRAM FOR b AND e.

It may be that the circumstances are such that no useful conclusions can be drawn from a diagram such as in Figure 42, but it should always be drawn as a check upon the diagram of Figure 40.

Considering Figures 40 and 42, I adopt .28 as a reasonable average value for b.

Average value of b.

Value of t.   Then from the equation

$$.6908(1-t) = .3563 + b$$

we find $1-t = .92$, $t = .08$ very approximately. Knowing $1-t$, we fill in line 37, which is the value of $(1-t)\,T$, or E.

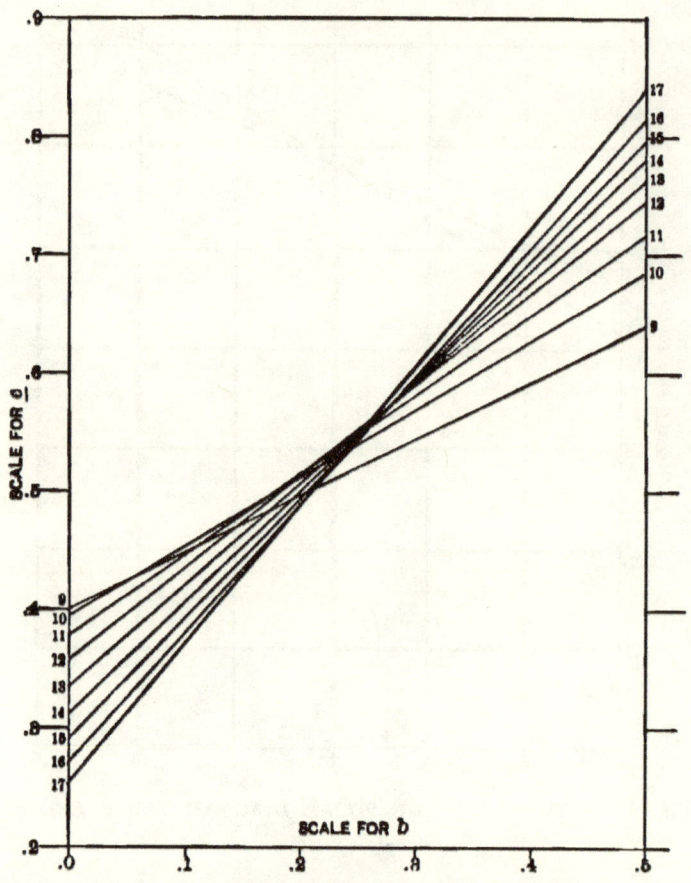

Fig. 42.— FOCAL DIAGRAM FOR b AND e.

Line 24. (repeating line 24) is then deducted from line 37, giving us in line 38 the values of E_w, or $.0030707\,bV^5$.

Values of b.   Dividing the quantities of line 38 by those of line 25 ($.0030707\,V^5$) we have the values of b, entered in line 39.

§ 3. YORKTOWN TRIAL ANALYSIS. 131

The eccentric variations of b at the lower speeds are due to the previously noted fact that at low speeds the values of E_w are so small, compared with the other absorbents of horse-power, that a small error in one of the latter causes a large error in the former. This is a matter of no practical importance, since the wave resistance at low speeds is so small as to be almost negligible.

Table X. is completed by entering in line 40 the values of the efficiency of propulsion. This is equal to $\frac{E}{I}$, but is most readily obtained by multiplying the values of the thrust efficiency in line 28 by $1-t$. *Efficiency of Propulsion.*

I need hardly say that the numerous multiplications and divisions of Table X. make a slide rule or some equivalent instrument almost a necessity, in order to avoid wearisome arithmetic. With a slide rule the whole work of analysis as shown by Table X., including the necessary focal diagrams, can easily be completed and checked in five or six hours. *Mechanical Calculation Necessary.*

The values of b and t above were necessarily obtained by a somewhat round-about process. Hence they cannot be regarded with the same confidence as the values of the wake factor w and the thrust coefficient a. It has been my experience, however, that if the methods described above are intelligently applied to fairly accurate trial results, the values of b and t deduced, while leaving much to be desired from the point of view of scientific accuracy, are quite close enough to the truth for practical purposes. It is an unfortunate fact that accurate trial results are rare — not only because of the rarity of full trials, but because even extended trials are sometimes painfully inaccurate. *Approximation of Results to Truth.*

A graphic account of the distribution of the indicated horse-power of the *Yorktown* will be found of interest. Figure 43 shows the amounts of the component horse-powers from 5 to 17 knots. *Distribution of Power at Various Speeds.*

The ordinates of the curve numbered 1 show the horse-power absorbed by initial friction. The load friction is given by the ordinates between 1 and 2. The ordinate between 2 and 4 shows the loss by friction and resistance of the propeller. The wake gain and thrust deduction loss are so

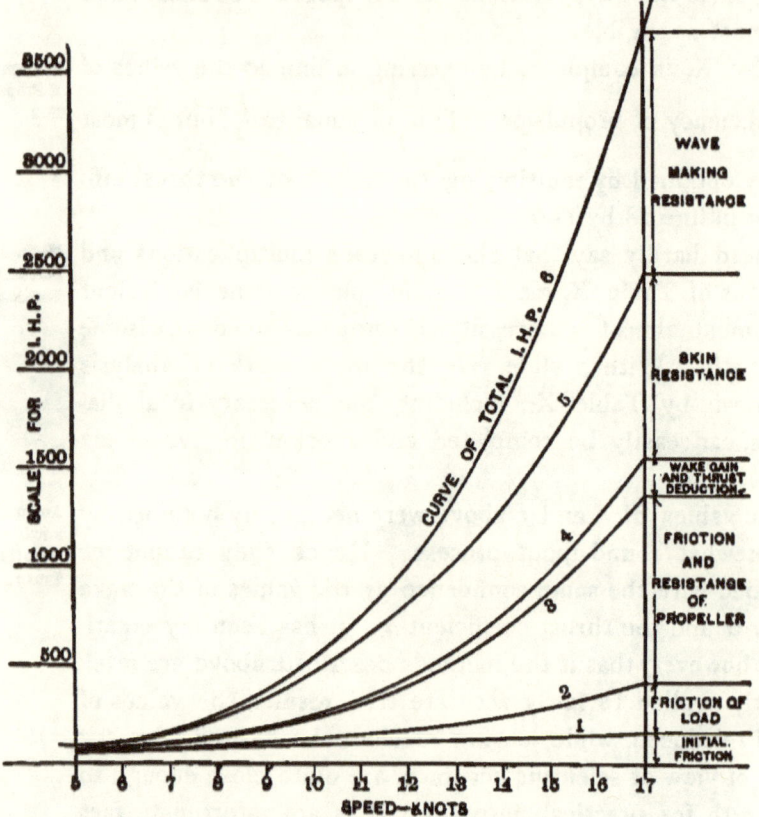

Fig. 43. — DISTRIBUTION OF YORKTOWN'S POWER.

nearly identical that they cannot be distinguished on the scale of the figure.

The wake gain is represented by the ordinate between 4 and 3, and the thrust deduction loss by the ordinate between 3 and 4.

The ordinate between 4 and 5 shows the skin resistance power, and the ordinate between 5 and 6 the wave resistance power. The ordinate of 6 measured from the base represents, of course, the total indicated power.

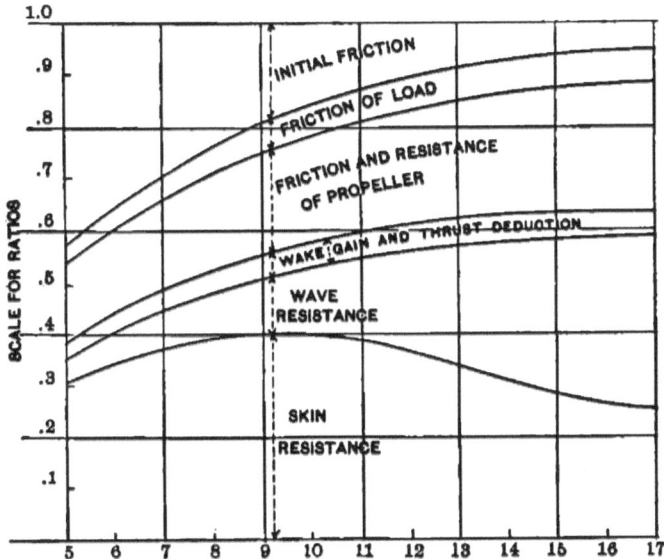

Fig. 44. — RATIOS IN WHICH YORKTOWN POWER IS DISTRIBUTED.

Figure 44 shows graphically the relative amounts of the various absorbents of the indicated power throughout the range of Figure 43. It calls for no detailed explanation. The large proportionate value of the initial friction at low speeds and the rapid increase in the importance of the wave resistance at the upper speeds are notable features. They are characteristic of all similar diagrams for fast ships.

§ 4. *Distribution of Power.*

It is interesting to compare the distribution of the *Yorktown* power at 16 knots with the average distribution as estimated by Froude some sixteen years ago, for the single-screw slow-speed merchant steamer of that day.

Distribution of Yorktown Power at 16 Knots.

For the *Yorktown* at 16 knots,

Initial friction power	$= .05$ I. H. P.
Friction of load	$= .07$ I. H. P.
Friction and resistance of propeller	$= .29$ I. H. P.
Wake gain	$= .05$ I. H. P.
Thrust deduction loss	$= .05$ I. H. P.
Effective H. P.	$= .59$ I. H. P.

Of the effective horse-power an amount $= .271$ of the I. H. P. is absorbed by skin resistance, and an amount $= .319$ of the I. H. P. by wave resistance.

Froude's Estimate.

Froude's average estimate was as follows:

Initial friction absorbs	.130 I. H. P.
Load friction absorbs	.130 I. H. P.
Air pump resistance, etc.,	.070 I. H. P.
Screw resistance,	.129 I. H. P.
Thrust deduction,	.154 I. H. P.
Effective horse-power,	.387 I. H. P.

Froude's screw resistance, as given above, is the actual screw resistance less the gain from the wake. If this be borne in mind, the ratios between the power absorbed by the screw and the effective horse-power will not be greatly different in Froude's estimate and the *Yorktown* analysis.

The proportionate loss by initial friction has been steadily decreasing for some years. This is due partly to better design and increasing excellence of workmanship, but more largely to the adoption of high pressures, resulting in a higher mean effective pressure.

If the initial friction were equivalent to 2 pounds per square inch upon the low-pressure piston area, and the mean effective pressure reduced to the low pressure was 15 pounds

§ 4.   DISTRIBUTION OF POWER.   135

for a boiler pressure of about 60 pounds, the power lost by initial friction would be $\frac{2}{15}$, or .133, of the total indicated. But if the boiler pressure be increased to 160 pounds, or more, involving a rise in the mean effective pressure to, say, 25 pounds, then the loss by initial friction will be but $\frac{2}{25}=.080$ of the total indicated power.

It is interesting to compare Froude's estimates with some recent estimates published by Mr. Blechynden in the transactions of the N. E. Coast Institute. Mr. Blechynden's estimates assume that all pumps are driven off the main engines. They apply more particularly to merchant vessels with modern triple-expansion engines. Mr. Blechynden's figures are as follows: *Blechynden's Estimates.*

Total power indicated,	100
Absorbed by dead load and air pump,	7.8
Absorbed by circulating pump,	1.5
Absorbed by feed pump,	0.6
Absorbed by bilge pump,	0.5
	10.4
Absorbed by load friction $=.075(100-10.4)$,	6.7
	17.1
Delivered to propeller,	82.9

If no pumps except air pumps are driven off the main engines, Mr. Blechynden's estimates become as below:

Total indicated,	100
Absorbed by dead load and air pump,	7.8
Absorbed by load friction,	6.9
	14.7
Delivered,	85.3

Mr. Blechynden states also that the initial friction and the power to work pumps absorb on the average from 1.5 to 1.75 pounds per square inch of effective pressure upon the total area of all pistons.

§ 5. *Coefficients and Constants for Practical Use.*

It is best when one has to estimate the performance of a ship or her engines to adopt coefficients obtained from machinery of the same, or nearly the same, type as that in hand — allowing always a fair margin for possible small errors and differences. Where no data from similar machinery is available, the following may serve as a guide:

Efficiency of Engine.

1. When no pumps are driven off the main engines, from 84 to 89 per cent of the indicated horse-power at the highest speed is delivered to the propeller, when the engines are of the triple-expansion type, and the boiler pressure is high.

The variation of the coefficient of delivery is due partly to differences in types of machinery and partly to variations in workmanship, adjustment of bearings, etc. For instance, an engine with plain slide valves worked by double eccentric gear would absorb more power in initial friction than an engine with piston valves throughout and radial valve gear.

2. When all pumps are driven off the main engines, from 80 to 85 per cent of the indicated horse-power should be delivered to the propeller.

The coefficient in this case also depends upon the workmanship and the type of machinery.

Value of b.

The coefficient b is, in the nature of things, a somewhat uncertain quantity. But since a fair approximation to the wave resistance is obtained by supposing it equal to bV^4, it is desirable to form some idea of the effect upon the value of b of change in size and type of vessel.

Since the wave resistance satisfies the law of comparison, it must, if it varies as V^4, vary also in passing from one

§ 5.   COEFFICIENTS AND CONSTANTS.   137

vessel to a similar one as the linear dimensions (see page 30). This is to say, the wave resistances of similar vessels of different sizes will be denoted by $b_0 V^4 \times$ (a quantity proportional to the linear dimensions of the vessel).

Now we can write any number of functions of L, B, H, and D, which will be proportional to the linear dimensions.

It is desirable to choose a simple function which changes but little, if at all, for change of type of vessel throughout the range of types of vessel ordinarily met with.

The most satisfactory function which I have been able to discover is $\dfrac{D^{\frac{2}{3}}}{L}$.

Adopting this function, it is found that a fair average value of b_0 for moderately fine high-speed vessels is .4. For vessels broad in proportion to their length, especially if the ends are so fine as to make the effective length less than the actual length, the value of b_0 rises to .45 or so. *Value of b_0.*

For vessels which are long in proportion to their beam, and at the same time moderately fine, such as the majority of the modern Atlantic liners, the value of b_0 will fall to .35 or so.

Of course the "lines" of a vessel, or the "model," so called, must influence the wave-making resistance. But so far as my experience goes, given the extreme dimensions of a vessel, and the displacement on those dimensions, any difference in lines which appears in practice has very little effect upon wave-making resistance. *Influence of Model on Lines.*

For full vessels, *i.e.* with block coefficients above .6, the value of b_0 appears to increase above .45. Such vessels, however, are seldom driven at high speeds, so that their wave resistance is proportionately small. In dealing with them, unless we have a reliable coefficient from a ship of similar type, it is well enough, as a rule, to take b_0 from .45 to .50.

Final Wave Resistance Formula.

Then the final formula for the wave resistance of a vessel is

$$R_w = b_0 \frac{D^{\frac{2}{3}}}{L} V^4,$$

where b_0 is a coefficient, usually about .4 for fast vessels, and rising to .5, or even more, for full and slow vessels.

In considering the processes by which the indicated horse-power is transmuted into the effective horse-power, it is best to divide the whole number of processes into three groups, so to speak.

Engine Efficiency.

1. In passing from the indicated horse-power I to the propeller power P there is a certain coefficient of transfer which may be called the engine efficiency and denoted by e.

Propeller Efficiency.

2. The power delivered to the propeller is transferred with a certain efficiency to the thrust block, the true efficiency being the propeller efficiency denoted by c_2.

Hull Efficiency.

3. There is a certain gain from the wake, and a certain loss by the thrust deduction, and net result being that the effective horse-power is simply that part of the thrust horse-power obtained from the indicated horse-power multiplied by the factor $\frac{1-t}{1-w}$.

This factor, dependent upon the hull, Froude called the hull efficiency, though hull coefficient would seem to be a more fitting designation. I shall denote it by c_3.

Then, $$E = c_1 c_2 c_3 \times I,$$

or $$e = \frac{E}{I} = c_1 c_2 c_3.$$

I have already sufficiently discussed c_1 and c_2. I propose now to say something about c_3, or rather about w and t, upon which c_3 depends.

§ 6. Thrust Deduction and Wake Factor.

The wake factor is a somewhat variable quantity. We have seen that the wake is made up of several components, of which the most important are the frictional wake and the stream-line wake. Since additional wake means gain of power, it would seem desirable at first sight to make the wake as great as possible.

Variation of Wake and Thrust Deduction.

We cannot change the frictional wake much, but by making the stern full we can increase the stream-line wake. Unfortunately when we fill out the lines toward the stern, we increase the thrust deduction, and so bring in an extra loss which counterbalances the gain from the additional wake.

Since increase of wake is accompanied by increase of thrust deduction, we may naturally expect the hull efficiency e_3, or $\frac{1-t}{1-w}$ to vary from ship to ship less than either t or w.

Froude, who is the leading authority upon this subject, states that the average value of the hull efficiency is unity. In design work, then, unless we have a good value of the hull efficiency from analysis of a trial of a vessel, similar to the design in hand, it is the best plan to take the hull efficiency as unity. It will be observed that for the *Yorktown* $w = .083$, $t = .08$, and the hull efficiency $= \frac{.92}{.917}$, which is very close to unity.

Hull Efficiency.

While the hull efficiency changes but little, the wake factor and the thrust deduction change a good deal from ship to ship. Figure 45 shows values of wake factor for various ships as given by Froude, plotted upon the block coefficients of fineness of the ships.

Wake Factors of Actual Ships.

It is to be noted that—

1. The wake factors of the single-screw ships are decidedly greater than those of the twin-screw ships.
2. The wake factors increase with the block coefficients.

Both facts agree with the theories of resistance, etc., which I have set forth.

In the single-screw ships the screw is more favourably situated to catch both the frictional wake and the stream-line wake, since both are at their maximum at the stern, in the centre line. Unfortunately, the single screw, because of its location, tends to produce greater thrust deduction than twin screws, so that the greater wake is not accompanied by greater hull efficiency.

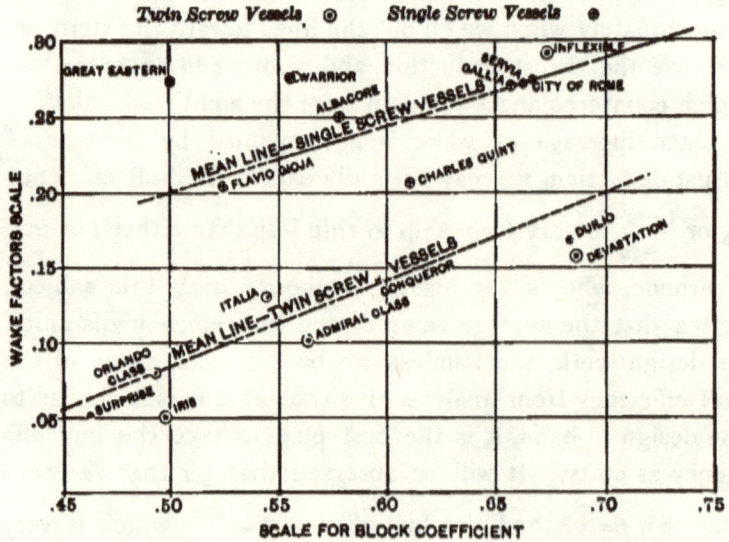

Fig. 45.—WAKE FACTORS PLOTTED ON BLOCK COEFFICIENTS.

The increase in wake with increase of block coefficient is due to the increase of stream-line wake due to the fulness aft accompanying increase of block coefficient. The abnormal wake of the *Inflexible*, for instance, is undoubtedly due to her full form aft.

Wake Factors for Design Work. For purposes of design it is always necessary to estimate the wake. This is best done from results of trials of similar ships.

§ 6. THRUST DEDUCTION AND WAKE FACTOR. 141

Those to whom such results are not available cannot do better than use the mean lines of wake of Figure 45. It is not necessary to know the wake very exactly, and the mean lines of Figure 45 will give results sufficiently near the truth in the vast majority of cases.

We shall see later that it is probably better to over-estimate the wake than to underestimate it, but that in either case we can adjust the blades of the propeller to suit the actual wake without any appreciable diminution of the efficiency.

CHAPTER V.

THE POWER OF SHIPS.

§ 1. *Preliminary.*

How are we to estimate with sufficient approximation the indicated horse-power which must be developed by the engines of a given ship in order to drive her at the intended speed?

Evidently this indicated horse-power will depend not only upon the effective horse-power, but also upon the efficiency of propulsion, or ratio between effective and indicated horse-power.

Methods divided into Two Classes. We may, in making estimates, deal with the indicated horse-power directly, or consider separately the quantities upon which it depends.

Accordingly the methods used fall naturally into two classes.

1. To the first class belong methods which deal at once with the indicated horse-power, without considering its components.

2. To the second class belong the methods which deal separately with the effective horse-power and the efficiency of propulsion.

Let us examine the leading methods in each class.

§ 2. *Admiralty Coefficient Method.*

Assumptions and Deduction Suppose that the resistance consists of skin resistance only,

§ 2. ADMIRALTY COEFFICIENT METHOD. 143

Then we may write

$$R = \frac{SV^2}{\text{constant}}.$$

If the efficiency of propulsion were constant, we should have

$$I = RV \times \frac{.0030707}{\text{constant efficiency}} = \frac{SV^3}{\text{constant}}.$$

For similar ships of varying sizes, S varies as the square of the linear dimensions. Now $D^{\frac{2}{3}}$ is proportional to the square of the linear dimensions, and so is the area of the midship section, which we denote by M.

So for similar ships with the same efficiency of propulsion we have

$$I = \frac{D^{\frac{2}{3}} V^3}{C_1} \text{ or } I = \frac{MV^3}{C_2},$$

where C_1 and C_2 are constant.

These formulæ may be taken to apply approximately to ships which are not similar, and whose efficiencies of propulsion differ.

C_1 and C_2 are then what are called the admiralty coefficients, C_1 being the "displacement coefficient," and C_2 the "midship section" coefficient. *Two Coefficients.*

Unfortunately the indicated horse-power of a ship does not vary as the cube of the speed, and hence these coefficients are not constant, even for the same ship at various speeds. *Variation of Coefficients with Speed.*

Figure 46 shows curves of the "coefficients" for the *Yorktown*, plotted upon speed. The variation is seen to be great, the values of the coefficients being least at the lowest and the highest speed, and reaching a maximum at about 10 knots.

The two curves are essentially the same, their ordinates bearing a constant ratio to each other.

For
$$C_1 = \frac{D^{\frac{2}{3}} V^3}{I},$$

$$C_2 = \frac{M V^3}{I},$$

whence
$$\frac{C_1}{C_2} = \frac{D^{\frac{2}{3}}}{M} = \text{a constant}.$$

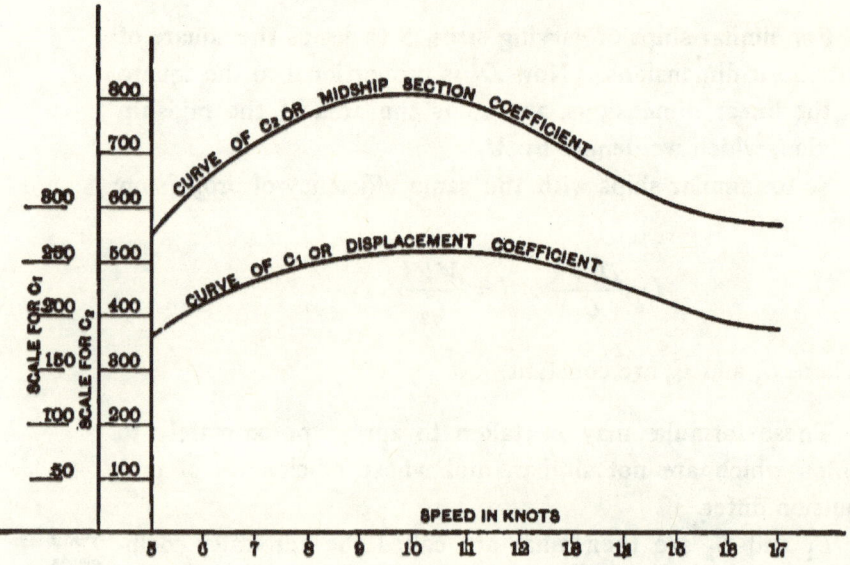

Fig. 46.—CURVES OF ADMIRALTY COEFFICIENTS FOR YORKTOWN.

The shape of curves of the coefficients shown in Figure 46 is typical for a fast ship. The reason why the shape should be as shown is not far to seek.

Cause of Variation with Speed.

We have
$$C_1 = \frac{D^{\frac{2}{3}} V^3}{I},$$

and
$$I = \frac{E}{e} = .0030707 \left(\frac{fSV^{2.83} + bV^5}{e} \right).$$

§ 2. ADMIRALTY COEFFICIENT METHOD. 145

Substituting and reducing

$$C_1 = \frac{D^{\frac{2}{3}}}{.0030707} \times \frac{V^3}{fSV^{2.83}+bV^5} \times e$$

$$= k\frac{V^3}{fSV^{2.83}+bV^5} \times e,$$

where k is constant for a given ship. Now at low speeds of a ship, bV^5 is comparatively small, and the expression $\frac{V^3}{fSV^{2.83}+bV^5}$ does not change much with change of speed.

But the value of e, the efficiency of propulsion, is increasing rapidly with the speed. So the value of C_1 increases at low speeds.

As the speed increases, we reach a point where

$$\frac{V^3}{fSV^{2.83}+bV^5}$$

has a maximum value, and then, owing to the rapid increase of the term bV^5 in the denominator, it begins to fall off, decreasing continually with the speed. Meanwhile e increases less rapidly as the speed increases, and at the highest speeds is usually nearly constant for several knots.

The result is that C_1 must reach a maximum value at some speed, and then constantly decrease with increase of speed.

Evidently, then, the admiralty coefficients for a fast ship and a slow ship may be expected to be very different, and a coefficient deduced from the performance of a slow ship will be unreliable for use in connection with a fast ship.

Also the coefficients may be expected to differ if the proportions of a ship be changed. This is because the wetted surface does not vary as $D^{\frac{2}{3}}$ except for precisely similar ships.

Variation of Coefficients with Proportions.

Thus, suppose we had a ship of 1000 tons' displacement and 6000 feet wetted surface; then $D^{\frac{2}{3}} = 100$ and $S = 60\ D^{\frac{2}{3}}$.

Now, retaining the same lines and the same midship section, let us double the length, and hence the displacement.

The wetted surface would be very nearly doubled, becoming 12,000.

D being now 2000, $D^{\frac{2}{3}} = 158.7$.

So we have now $S = 76\, D^{\frac{2}{3}}$ nearly, a change of about 25 per cent in the ratio between S and $D^{\frac{2}{3}}$.

With the midship section coefficient, C_2, the case would be even worse. The midship section of the 2000-ton ship being the same as that of the 1000-ton ship, the ratio $\dfrac{S}{M}$ for the 2000-ton ship would be double its value for the 1000-ton ship.

Evidently, then, the admiralty coefficients of ships which differ much in their proportions should be expected to be different.

Limitations of Method.
So when using this method to approximate to the power of a new design, we can rely only upon coefficients obtained from the performance of ships of similar proportions at similar speeds.

When the admiralty coefficients were introduced many years ago, there were few high-powered ships, and the wave resistance was seldom so great that a curve of coefficients would fall much below its maximum at the highest speed of the ship.

Present Status of Method.
The introduction of high powers and the multiplication of types were accompanied by such variations in the admiralty coefficients from ship to ship, that this method was soon discredited, and is now little used in England.

The French engineers still use the reciprocal of the midship-section coefficient with success; but they fully understand its restrictions, and in applying it to a new design are guided by results from ships as nearly as possible similar to the new design.

Their success is due to the possession of data from careful trials of numerous types of vessels, and to their own skill and experience in handling what is essentially a treacherous and untrustworthy method.

§ 3. *Kirk's Analysis.*

The fundamental idea underlying the admiralty coefficient method is that the resistance consists almost entirely of skin resistance, varying as the square of the speed. <small>Same Fundamental Idea as Preceding.</small>

We saw that one weakness of the method was due to the fact that the wetted surface does not vary as $D^{\frac{2}{3}}$, nor as M, except for similar ships. We may then naturally expect to find a method based upon the same fundamental idea, but using the wetted surface itself, or a reliable approximation to it.

This is the method brought forward by Dr. A. C. Kirk. The method is usually called Kirk's Analysis, and may be divided into two parts.

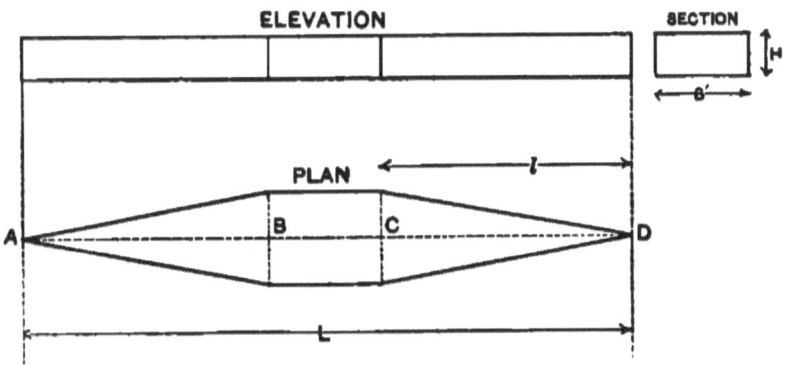

Fig. 47. — BLOCK MODEL OF YORKTOWN. SCALE: $1'' = 80'$.

The first part of Kirk's analysis is a method for approximating to the wetted surface.

The actual hull below the water line is represented by a "block model," or model with plane sides.

Let L denote the length of a ship, B the beam, H the mean draught, D the displacement in tons of 35 cubic feet, and M the area of the midship section. <small>Block Model.</small>

The block model, Figure 47, has the same length as the ship, and a uniform draught equal to the mean draught of the ship. The beam of the block model, denoted by B', is such that $B'H = M$.

Referring to Figure 47, let l denote the length CD, or AB, the length of the triangular entrance and run. CD and AB, or l, are made of such a length that the displacement of the block model is the same as that of the ship.

Then we have the following relations between the known dimensions, etc., of the ship and the dimensions of the block model:

$$B'H \times BC + 2 \times \tfrac{1}{2} B'H \times CD = D \times 35, \text{ or } M(L - 2l) + Ml = 35 D;$$

$$M(L - l) = 35 D;$$

$$l = L - \frac{35 D}{M}.$$

Surface of Block Model.
The surface of the block model is evidently nearly the same as that of the corresponding ship.

We have

Surface of bottom $= B'(L - l) = \dfrac{M}{H} \cdot \dfrac{35 D}{M} = \dfrac{35 D}{H}$.

Surface of parallel sides $= 2 H(L - 2l)$.

Surface of sides of ends $= 4 H \sqrt{\left(\dfrac{B'}{2}\right)^2 + l^2}$.

The total block model surface is readily calculated from the above. It is as a rule greater than the actual wetted surface. Kirk estimates the excess as follows:

For very full ships the excess is about 2 per cent.

For ordinary merchant steamers the excess is about 3 per cent.

For fine steamers, but not with hollow water lines, the excess is about 5 per cent.

For very fine steamers with hollow water lines, the excess is about 8 per cent.

Formula of Kirk's Analysis.
Having determined the wetted surface S, as above, with sufficient approximation, Kirk assumed that the resistance

§ 3. KIRK'S ANALYSIS.

varied directly as the surface and as the square of the speed, and that the indicated horse-power would vary as the cube of the speed. Then he adopted the formula

$$I = \frac{kSV^3}{100000},$$

where k is a coefficient.

According to Kirk, the appropriate values of k under various conditions are as below:

1. For fine ships with smooth and clean bottom and high efficiency of propulsion, $k=4$.   *Values of Constant.*
2. For merchant ships of ordinary proportions and efficiency, $k=5$.
3. For short, broad ships, $k=6$.

The above formula was proposed for and intended to be applied to ships of moderate speed, — not above 12 knots or so, — being more especially adapted to merchant vessels of 10 knots' speed.   *Limitations and Errors of Method.*

It has one advantage over the admiralty coefficient method, in that it uses a close approximation to the wetted surface itself instead of quantities whose ratio to the wetted surface varies with variation in the proportions of ships.

Hence k may be expected to change less from ship to ship than C_1 and C_2, *ante*. But owing to the fact that the formula takes no account of the change in efficiency of propulsion in passing from low to high speeds, and no adequate account of the wave resistance, it is found that for a given ship, k varies greatly with the speed.

This fact is exemplified by Figure 48, which is a curve of values of k for the *Yorktown*.

It may be noted that the values of k and of the admiralty coefficients are connected by a reciprocal relation.   *Connection between k and Admiralty Coefficients.*

Thus, $$I = \frac{D^{\frac{2}{3}} V^3}{C_1} = \frac{kSV^3}{100000},$$

whence, $$C_1 k = \frac{100000\, D^{\frac{2}{3}}}{S} = \text{a constant for a given ship.}$$

Efficiencies implied by Kirk's Constants.

Let us see what the values of k given by Kirk imply as regards efficiency.

For a steamer, say 350 feet long, the coefficient of skin friction is, from Table V., .00916. From Table XI. the value of $10^{1.83}$ is 67.61.

Hence the skin resistance at 10 knots of 100 square feet of wetted surface of such a steamer is $100 \times .00916 \times 67.61 = 62$ pounds nearly.

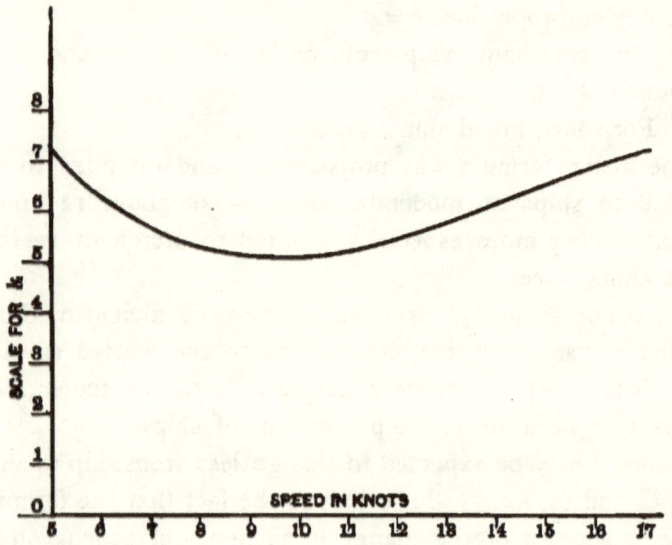

Fig. 48. — CURVE OF KIRK'S CONSTANT k FOR YORKTOWN.

The skin resistance power corresponding is

$$.0030707 \times 10 \times 62 = 1.9.$$

The corresponding indicated horse-power by Kirk's analysis is $\dfrac{k \times 100 \times 1000}{100000} = k.$

If $k = 4$, the efficiency of propulsion, considering skin resistance alone, is $\dfrac{1.9}{4} = .475.$

§ 4.   EXTENDED LAW OF COMPARISON.   151

Similarly, the efficiencies corresponding to $k=5$ and $k=6$ are .38 and .317 respectively.

These numbers agree very well with what might have been anticipated from our previous work.

While Kirk's analysis is preferable to the admiralty coefficient method, it is based upon the same assumptions, which, though fairly close approximations at low speeds, are seriously in error at high speeds. Hence the method should always be used with caution. In applying it to a new design, values of the constant k should be used which have been deduced from careful trials of actual ships similar in type and speed to the new design. *Value of Kirk's Analysis.*

§ 4. *Extended Law of Comparison.*

We have seen that, using the law of comparison, if we know for a given range of speed of a model or small ship the resistance following the law, we can determine the corresponding resistance of a full-sized ship, or a similar larger ship, throughout the corresponding range of speed.

The question naturally arises whether the Law of Comparison cannot be extended to indicated horse-power, even if only in an approximate manner. *Deduction of Extended Law.*

Froude has shown that his law can readily be applied to power as well as resistance, and a method of estimating indicated horse-power based upon the law of comparison is already much used and is rapidly growing in favour.

Suppose that all the resistance of ships followed the Law of Comparison. Let R, V, D, R_1, V_1, D_1 denote the resistances, corresponding speeds, and displacements of two similar ships.

Then
$$\frac{R}{R_1} = \frac{D}{D_1} \text{ by the law,}$$

and
$$\frac{V}{V_1} = \sqrt{\frac{L}{L_1}} = \sqrt{\left(\frac{D}{D_1}\right)^{\frac{1}{3}}} = \left(\frac{D}{D_1}\right)^{\frac{1}{6}}.$$

Hence, $\dfrac{RV}{R_1V_1} = \dfrac{D}{D_1} \times \left(\dfrac{D}{D_1}\right)^{\frac{1}{3}} = \left(\dfrac{D}{D_1}\right)^{\frac{4}{3}}.$

Now R, V and R_1, V_1 are proportional to E and E_1, the effective horse-powers.

Whence, $\dfrac{E}{E_1} = \dfrac{RV}{R_1V_1} = \left(\dfrac{D}{D_1}\right)^{\frac{4}{3}}.$

Now $E = cI$, $E_1 = e_1 I_1$, e and e_1 being the efficiencies of propulsion.

If e and e_1 are the same, which will be approximately the case in good work, we shall have

$$\dfrac{I}{I_1} = \dfrac{E}{E_1} = \left(\dfrac{D}{D_1}\right)^{\frac{4}{3}},$$

or $\quad I_1 = \left(\dfrac{D_1}{D}\right)^{\frac{4}{3}} I,$

and $\quad V_1 = \left(\dfrac{D_1}{D}\right)^{\frac{1}{3}} V.$

Hence, knowing I, V, and D for a given ship, we can readily calculate I_1 and V_1 for a similar ship of known displacement D_1.

So from a speed curve of a ship of a given size we can deduce a corresponding curve for a similar ship of any size.

Example of Derived Power Curves. Figure 49 shows the actual speed curve of the *Yorktown* at 1680 tons extending to 17 knots, and the corresponding curves deduced by the above method for similar ships of 1000 tons and 2500 tons, extending to 15.6 and 18.2 knots respectively.

Errors of Extended Law. There are four principal sources of error in using the Extended Law of Comparison.

In the first place, as we know, the skin resistance does not exactly follow the law. Take for instance the case of the *Yorktown*.

§ 4.   EXTENDED LAW OF COMPARISON.   153

The skin resistance power of the *Yorktown* at 16 knots was 800. **Skin Resistance does not follow Law.**

For a 10,000-ton vessel similar to the *Yorktown*

$$\left(\frac{D_1}{D}\right)^{\frac{1}{6}} = \left(\frac{10000}{1680}\right)^{\frac{1}{6}} = 1.346 \text{ (the speed factor)},$$

while the power factor $= \frac{10000}{1680} \times 1.346 = 8.013$.

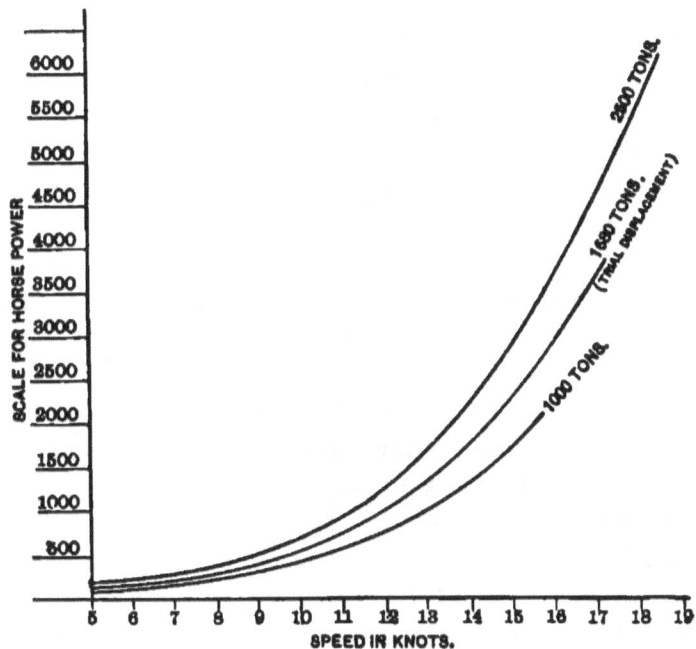

Fig. 49. — POWER CURVE OF YORKTOWN AT 1680 TONS. REDUCED TO 1000 AND 2500 TONS.

Then if the law applied, the skin resistance power of the 10,000-ton ship at $(16 \times 1.346) = 21.54$ knots would be $(800 \times 8.013) = 6410$.

Now the actual wetted surface of the *Yorktown* at 1680 tons was 10,840 square feet.

The wetted surface of a similar 10,000-ton vessel would be 10,840 $(\frac{10000}{1680})^{\frac{2}{3}} = 35,610$ square feet.

The length of the *Yorktown* being 226 feet, that of the similar 10,000-ton ship would be $(\frac{10000}{1680})^{\frac{1}{3}} \times 226 = 409.6$ feet.

The appropriate coefficient of friction (Table V.) is .0091. Then this skin resistance power would be $35,610 \times .0091 \times .0030707 \times (21.54)^{2.83} = .995 \times 21.54^{2.83} = .995 \times 5930 = 5900$.

The Extended Law of Comparison gave 6410 for the skin resistance power. It is seen, then, that the extended law will tend to make us overestimate the power for a ship when working from results of a smaller ship.

Efficiency of Propulsion varies for Given Ship.

The second source of error in applying the extended law arises from the fact that the efficiency of propulsion is not constant — usually increasing in passing from a lower to a higher speed.

To illustrate: Suppose we wish to estimate the power necessary to drive a 10,000-ton ship at 9.42 knots. The corresponding speed of the *Yorktown* is 7 knots, at which 205 I. H. P. is developed.

The power factor to reduce the *Yorktown* to 10,000 tons being 8.013, we have for the power of the 10,000-ton vessel at 9.42 knots, $205 \times 8.013 = 1643$ I. H. P.

But this result, as appears from the deduction of the extended law, is obtained on the assumption that the efficiency of propulsion of the 10,000-ton ship at 9.42 knots is the same as that of the *Yorktown* at 7 knots.

At 7 knots the *Yorktown* efficiency is low, only about .45; while, since 9.42 is the top speed of the 10,000-ton vessel, we may reasonably expect her to show a much higher efficiency, say .54.

So, taking this into account, the true power for the 10,000-ton 9.42-knot vessel should be not 1642, but

$$1642 \times \frac{.45}{.54} = 1368 \text{ I. H. P.}$$

It is seen that this source of error also tends to make us overestimate the necessary power for a given ship when working from the performance of a smaller ship.

But since the efficiency of propulsion, especially in efficient ships, usually changes but little for several knots below the top speed, we may safely apply the extended law to the last three or four knots of a speed curve of an actual ship.

The two sources of error already discussed are in a sense sources of safety, since they tend to make us overestimate the necessary power.

The third source is common to all methods. It is due to the fact that we must sometimes use trial results of a ship of given proportions in estimates dealing with a ship of rather different proportions. *Extended Law does not allow for Variation of Type.*

With the exercise of sufficient care and a certain amount of judgment, the error from this source should be small.

The final source of error in using the extended law is due to the fact that the efficiency of propulsion of a new design upon completion may be greater or less than that of the ship, or ships from whose results we are working. *Efficiency of Propulsion of Ships differ.*

This error also should be small if the person making the estimate exercises care.

§ 5. *Standard Curves of Power.*

The application of the Extended Law of Comparison is much facilitated by the adoption of a "standard" displacement to which all trial data may be reduced.

A convenient standard displacement is 10,000 tons. In order to facilitate the work of standardising I have calculated Tables XII. and XIII. Table XII. gives factors by which the speed and power of ships of any size likely to occur in practice, must be multiplied in order to reduce them to the standard displacement of 10,000 tons. *Tables for Standardising.*

Table XIII. gives the factors by which the dimensions of

a given ship must be multiplied in order to obtain the corresponding dimensions of a similar ship of 10,000 tons' displacement.

I wish to call attention now to the standard speed and power curves shown by Figures 50 to 61. Each figure covers a range of one knot. Where the maximum speed only of a ship was available, it has been "spotted" in its proper place.

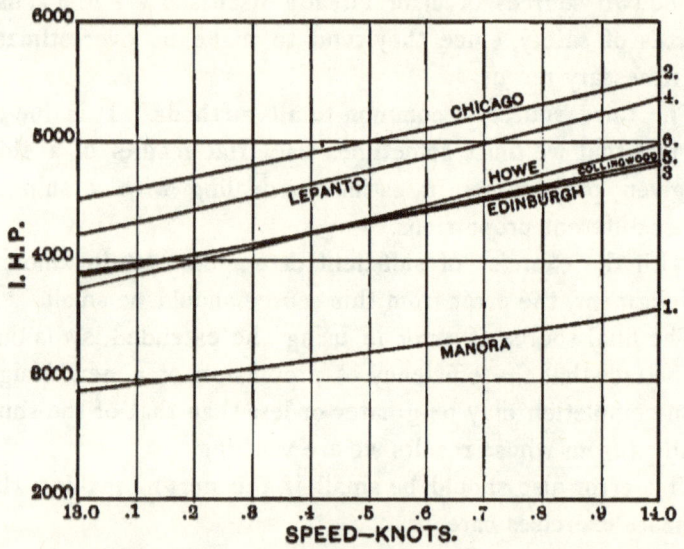

Fig. 50.— STANDARD CURVES. 13 TO 14 KNOTS.

Standard Power Curves. Facing each figure is a table giving sufficient information about the ships concerned, to enable any one using the diagrams to decide which vessels approach within reasonable limits, the size and type for which he wishes to determine the necessary power.

Where a complete speed curve was available, only the upper portion has been used for the standard diagrams in order to avoid errors due to the falling off of the efficiency of propulsion at the lower speeds.

The data used in preparing Figures 50 to 61 is not new, having been taken from various sources, such as papers before

DIMENSIONS OF SHIPS FOR STANDARD CURVES 13 TO 14 KNOTS.

Number.	Name.	At Trial Displacement.						At 10,000 Tons.			
		Length L. Feet.	Beam B. Feet.	Mean Draught H. Feet.	Displacement D. Tons.	Block Coefficients.	Top Speed V. Knots.	Length L. Feet.	Beam B. Feet.	Mean Draught H. Feet.	Top Speed V. Knots.
1	Manora	410	45.0	16.21	5120	·599	15.5	512	56.3	20.26	17.4
2	Chicago .	315	48.2	19.04	4543	·550	15.3	410	62.7	24.77	17.5
3	Edinburgh	325	68.0	22.71	7710	·533	16.0	354	74.2	24.77	16.7
4	Lepanto .	400	73.9	30.30	14860	·580	18.4	351	64.8	26.58	17.2
5	Collingwood	325	68.0	23.75	8200	·547	16.6	347	72.7	25.37	17.2
6	Howe	325	68.0	26.67	9637	·572	16.9	329	68.9	27.00	17.0

technical societies and accounts of trials. While I cannot vouch for its accuracy, I have used nothing that did not appear to be sufficiently accurate for practical purposes. Ample blank space has been purposely left so that any one possessing reliable data may add to his diagrams.

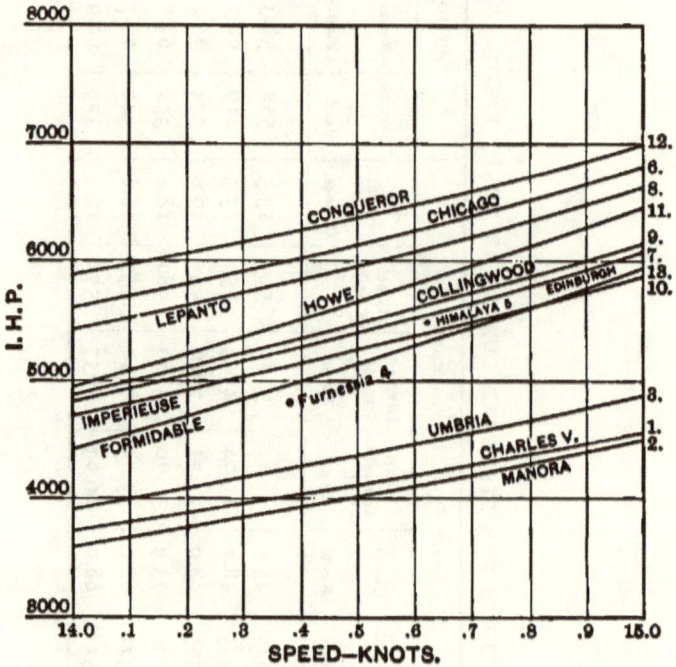

Fig. 51.—STANDARD CURVES. 14 TO 15 KNOTS.

Example of Use of Standard Diagrams. To illustrate the use of the diagrams, let us suppose that we wish to determine the necessary indicated horse-power for a ship of the following characteristics:

        Length,   279 feet.
        Beam,   39.2 feet.
        Mean draught,   16.40 feet.
        Displacement,   2500 tons.
        Block coefficient, .488
        Speed,   17 knots.

STANDARD CURVES OF POWER.

DIMENSIONS OF SHIPS FOR STANDARD CURVES 14 TO 15 KNOTS.

Number	Name	At Trial Displacement.						At 10,000 Tons.			
		Length L. Feet.	Beam B. Feet.	Mean Draught H. Feet.	Displacement D. Tons.	Block Coefficients.	Top Speed V. Knots.	Length L. Feet.	Beam B. Feet.	Mean Draught H. Feet.	Top Speed V. Knots.
1	Charles V.	323	33.0	14.83	2479	.571	15.0	514	52.5	23.25	18.9
2	Manora	410	45.0	16.21	5120	.599	15.5	512	56.3	20.26	17.4
3	Umbria	500	57.0	21.33	9860	.568	20.2	502	57.3	21.40	20.2
4	Furnessia	445	44.5	22.20	8578	.683	14.0	469	46.9	23.38	14.4
5	Himalaya	341	46.2	18.83	3857	.482	12.5	468	63.4	25.87	14.7
6	Chicago	315	48.2	19.04	4543	.550	15.3	410	62.7	24.77	17.5
7	Edinburgh	325	68.0	22.71	7710	.533	16.0	354	74.2	24.77	16.7
8	Lepanto	400	73.9	30.30	14860	.580	18.4	351	64.8	26.58	17.2
9	Collingwood	325	68.0	23.75	8200	.547	16.6	347	72.7	25.37	17.2
10	Imperieuse	315	61.0	24.90	7573	.554	17.2	346	66.8	27.32	18.0
11	Howe	325	68.0	26.67	9637	.572	16.9	329	68.9	27.00	17.0
12	Conqueror	270	58.0	22.56	6040	.599	15.5	319	68.6	26.69	16.9
13	Formidable	322	69.7	26.41	11220	.663	16.2	309	67.0	25.40	15.9

160   RESISTANCE OF SHIPS.   § 5.

For reducing to 10,000 tons we have the dimension factor (Table XIII.) = 1.587, and the speed factor (Table XII.) = 1.260.

Then upon standardising, we have

    Length,   443 feet.
    Beam,   62.2 feet.
    Mean draught,   26.02 feet.
    Speed,   21.42 knots.

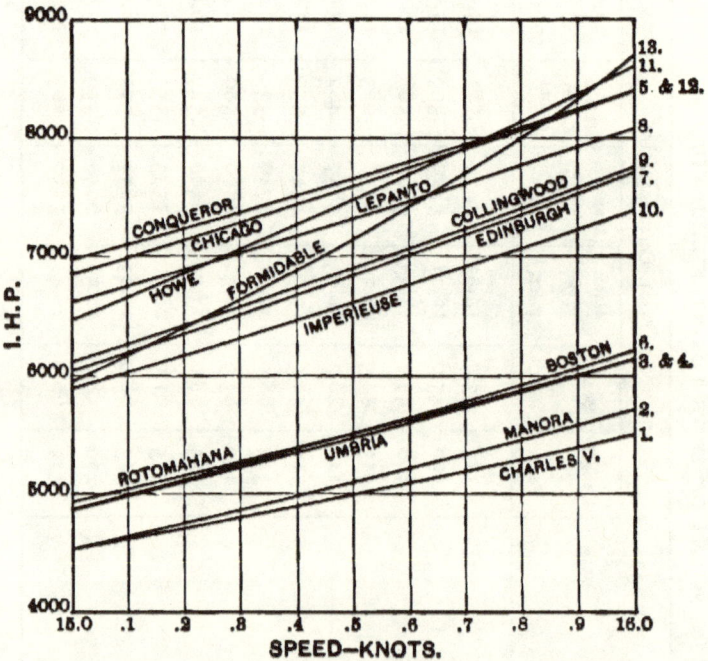

Fig. 52.—STANDARD CURVES. 15 TO 16 KNOTS.

Turning to Figure 58, which extends from 21 to 22 knots, we see that the powers required for 21.42 knots vary a good deal. The ships most nearly resembling the one in hand appear to be the *Condor* and *Faucon*, the *Iris* and the *Curlew*.

§ 5. STANDARD CURVES OF POWER. 161

DIMENSIONS OF SHIPS FOR STANDARD CURVES 15 TO 18 KNOTS.

Number	Name	At Trial Displacement.						At 10,000 Tons.			
		Length L. Feet.	Beam B. Feet.	Mean Draught H. Feet.	Displacement D. Tons.	Block Coefficients.	Top Speed V. Knots.	Length L. Feet.	Beam B. Feet.	Mean Draught H. Feet.	Top Speed V. Knots.
1	Charles V.	323	33.0	14.83	2479	.571	15.0	514	52.5	23.25	18.9
2	Manora	410	45.0	16.21	5120	.599	15.5	512	56.3	20.26	17.4
3	Umbria	500	57.0	21.33	9860	.568	20.2	502	57.3	21.40	20.2
4	Rotomahana	285	35.0	14.93	2425	.569	15.4	457	56.1	23.90	19.5
5	Chicago	315	48.2	19.04	4543	.550	15.3	410	62.7	24.77	17.5
6	Boston	270	42.0	17.66	3235	.564	15.6	394	61.2	25.73	18.8
7	Edinburgh	325	68.0	22.71	7710	.533	16.0	354	74.2	24.77	16.7
8	Lepanto	400	73.9	30.30	14860	.580	18.4	351	64.8	26.58	17.2
9	Collingwood	325	68.0	23.75	8200	.547	16.6	347	72.7	25.37	17.2
10	Imperieuse	315	61.0	24.90	7573	.554	17.2	346	66.9	27.32	18.0
11	Howe	325	68.0	26.67	9637	.572	16.9	329	68.8	27.00	17.0
12	Conqueror	270	58.0	22.56	6040	.599	15.5	319	68.6	26.69	16.9
13	Formidable	322	69.7	26.41	11220	.663	16.2	309	67.0	25.40	15.9

All things considered, it appears that a suitable standard power for the case in hand is 22,500. The power factor

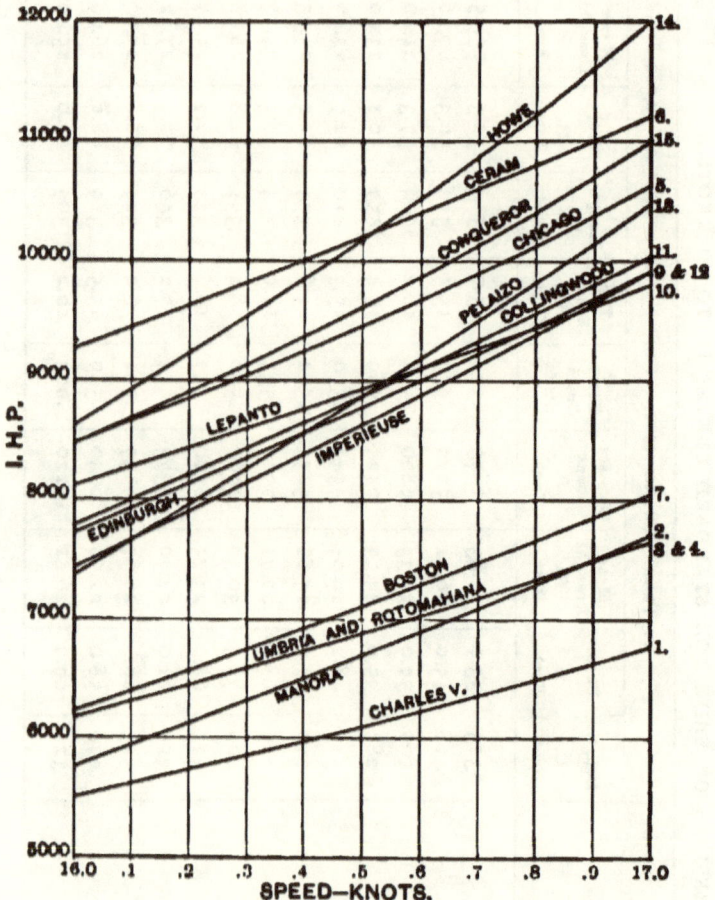

Fig. 53. — STANDARD CURVES. 16 TO 17 KNOTS.

from Table XII. being 5.040, the actual power for the 2500-ton ship at 17 knots, as deduced by this method, will be $\frac{22500}{5.040} = 4465$.

STANDARD CURVES OF POWER.

DIMENSIONS OF SHIPS FOR STANDARD CURVES 16 TO 17 KNOTS.

Number	Name	At Trial Displacement.						At 10,000 Tons.			
		Length L. Feet.	Beam B. Feet.	Mean Draught H. Feet.	Displacement D. Tons.	Block Coefficients.	Top Speed V. Knots.	Length L. Feet.	Beam B. Feet.	Mean Draught H. Feet.	Top Speed V. Knots.
1	Charles V.	323	33.0	14.83	2479	.571	15.0	514	52.5	23.25	18.9
2	Manora	410	45.0	16.21	5120	.599	15.5	512	56.3	20.26	17.4
3	Umbria	500	57.0	21.33	9860	.568	20.2	502	57.3	21.40	20.2
4	Rotomahana	285	35.0	14.93	2425	.569	15.4	457	56.1	23.90	19.5
5	Chicago	315	48.2	19.04	4543	.550	15.3	410	62.7	24.77	17.5
6	Ceram	152	25.6	9.45	512	.487	12.8	409	69.0	25.45	21.0
7	Boston	270	42.0	17.66	3235	.564	15.6	394	61.2	25.73	18.8
9	Edinburgh	325	68.0	22.71	7710	.533	16.0	354	74.2	24.77	16.7
10	Lepanto	400	73.9	30.30	14860	.580	18.4	351	64.8	26.58	17.2
11	Collingwood	325	68.0	23.75	8200	.547	16.6	347	72.7	25.37	17.2
12	Imperieuse	315	61.0	24.90	7573	.554	17.2	346	66.9	27.32	18.0
13	Pelayo	335	66.3	24.11	9745	.637	16.7	337	66.8	24.30	16.8
14	Howe	325	68.0	26.67	9637	.572	16.9	329	68.8	27.00	17.0
15	Conqueror	270	58.0	22.56	6040	.599	15.5	319	68.6	26.69	16.9

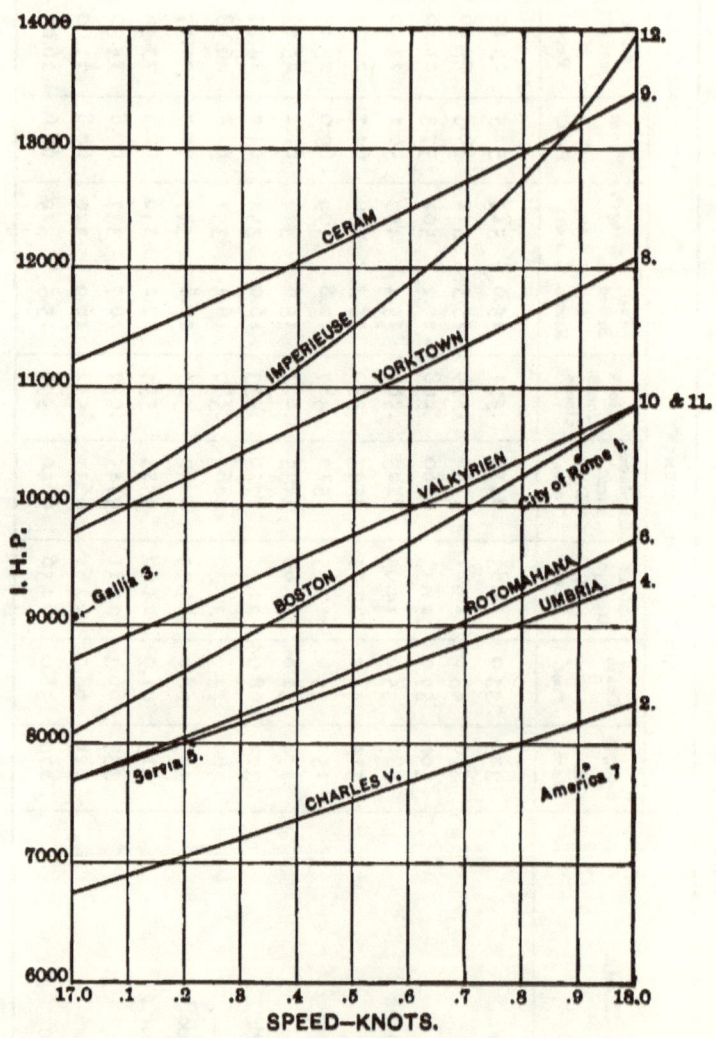

Fig. 54.—STANDARD CURVES. 17 TO 18 KNOTS.

§ 5. STANDARD CURVES OF POWER. 165

DIMENSIONS OF SHIPS FOR STANDARD CURVES 17 TO 18 KNOTS.

Number.	Name	At Trial Displacement.					At 10,000 Tons.				
		Length L. Feet.	Beam B. Feet.	Mean Draught H. Feet.	Displacement D. Tons.	Block Coefficients.	Top Speed V. Knots.	Length L. Feet.	Beam B. Feet.	Mean Draught H. Feet.	Top Speed V. Knots.
1	City of Rome	543	52.0	21.46	11230	.649	18.2	522	50.0	20.65	17.9
2	Charles V.	323	33.0	14.83	2479	.571	15.0	514	52.5	23.25	18.9
3	Gallia	430	44.0	18.42	6125	.615	15.7	506	51.8	21.69	17.0
4	Umbria	500	57.0	21.33	9860	.568	20.2	502	57.3	21.40	20.2
5	Servia	515	52.0	25.17	12450	.646	17.9	479	48.3	23.40	17.2
6	Rotomahana	285	35.0	14.93	2425	.569	15.4	457	56.1	23.90	19.5
7	America	432	51.0	26.58	9550	.571	17.8	439	51.8	26.98	17.9
8	Yorktown	226	36.0	13.84	1680	.523	16.7	410	65.2	25.08	22.4
9	Ceram	152	25.6	9.45	512	.487	12.8	409	69.0	25.45	21.0
10	Valkyrien	267	43.3	17.50	2923	.507	17.5	402	65.2	26.37	21.5
11	Boston	270	42.0	17.66	3235	.564	15.6	394	61.2	25.73	18.8
12	Imperieuse	315	61.0	24.90	7573	.554	17.2	346	66.9	27.32	18.0

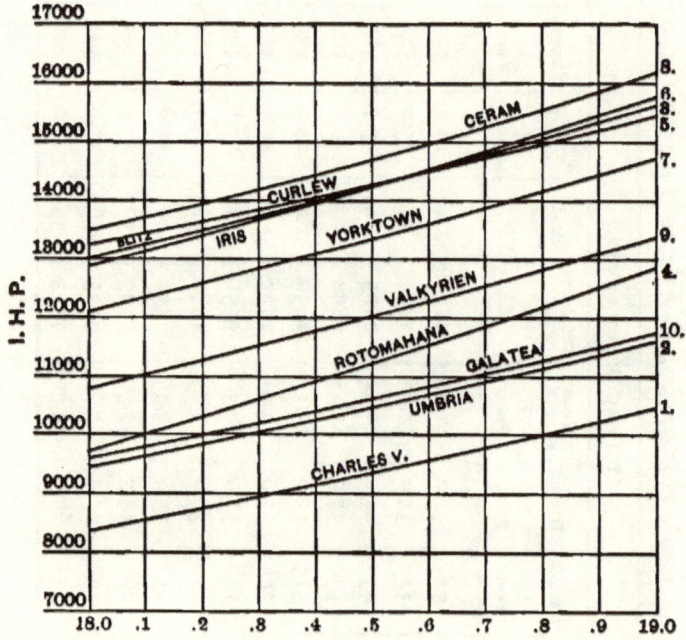

Fig. 55.—STANDARD CURVES. 18 TO 19 KNOTS.

§ 5. STANDARD CURVES OF POWER. 167

DIMENSIONS OF SHIPS FOR STANDARD CURVES 18 TO 19 KNOTS.

Number.	Name	At Trial Displacement.						At 10,000 Tons.			
		Length L. Feet.	Beam B. Feet.	Mean Draught H. Feet.	Displacement D. Tons.	Block Coefficients.	Top Speed V. Knots.	Length L. Feet.	Beam B. Feet.	Mean Draught H. Feet.	Top Speed V. Knots.
1	Charles V.	323	33.0	14.83	2479	.571	15.0	514	52.5	23.25	18.9
2	Umbria	500	57.0	21.33	9860	.568	20.2	502	57.3	21.40	20.2
3	Blitz	245	32.5	13.29	1382	.457	16.2	474	62.9	25.75	22.5
4	Rotomahana	285	35.0	14.93	2425	.569	15.4	457	56.1	23.90	19.5
5	Curlew	195	28.0	11.00	785	.458	14.5	455	65.4	20.70	22.2
6	Iris	300	46.1	17.08	3290	.488	18.6	435	66.7	24.74	22.4
7	Yorktown	226	36.0	13.84	1680	.523	16.7	410	65.2	25.08	22.4
8	Ceram	152	25.6	9.45	512	.487	12.8	409	69.0	25.45	21.0
9	Valkyrien	267	43.3	17.50	2923	.507	17.5	402	65.2	26.37	21.5
10	Galatea	300	56.0	21.00	5000	.496	19.0	378	70.6	26.46	21.3

168 RESISTANCE OF SHIPS. § 5.

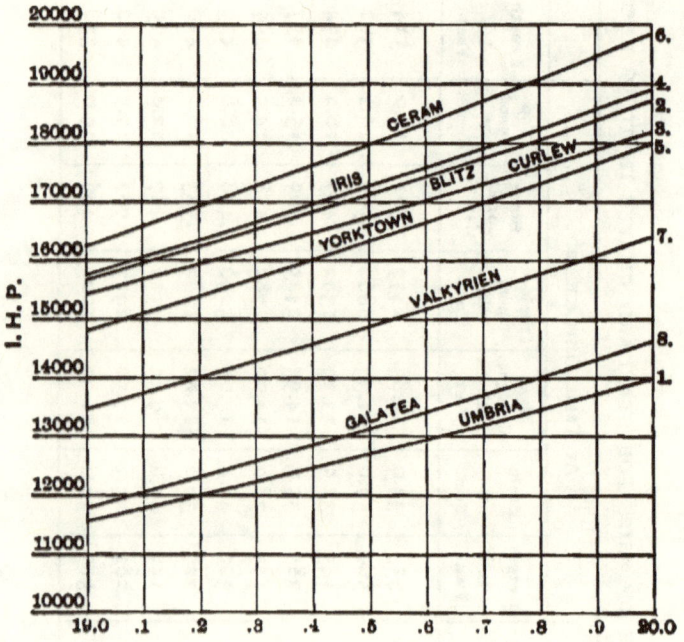

Fig. 56.—STANDARD CURVES. 19 TO 20 KNOTS.

§ 5.   STANDARD CURVES OF POWER.   169

DIMENSIONS OF SHIPS FOR STANDARD CURVES 12 TO 20 KNOTS.

Number	Name	At Trial Displacement.						At 10,000 Tons.			
		Length L. Feet.	Beam B. Feet.	Mean Draught H. Feet.	Displacement D. Tons.	Block Coefficients.	Top Speed V. Knots.	Length L. Feet.	Beam B. Feet.	Mean Draught H. Feet.	Top Speed V. Knots.
1	Umbria	500	57.0	21.33	9860	.568	20.2	502	57.3	21.40	20.2
2	Blitz	245	32.5	13.29	1382	.457	16.2	474	62.9	25.75	22.5
3	Curlew	195	28.0	11.00	785	.458	14.5	455	65.4	20.70	22.2
4	Iris	300	46.1	17.08	3290	.488	18.6	435	66.7	24.74	22.4
5	Yorktown	226	36.0	13.84	1680	.523	16.7	410	65.2	25.08	22.4
6	Ceram	152	25.6	9.45	512	.487	12.8	409	69.0	25.45	21.0
7	Valkyrien	267	43.3	17.50	2923	.507	17.5	402	65.2	26.37	21.5
8	Galatea	300	56.0	21.00	5000	.496	19.0	378	70.6	26.46	21.3

170 RESISTANCE OF SHIPS. §5.

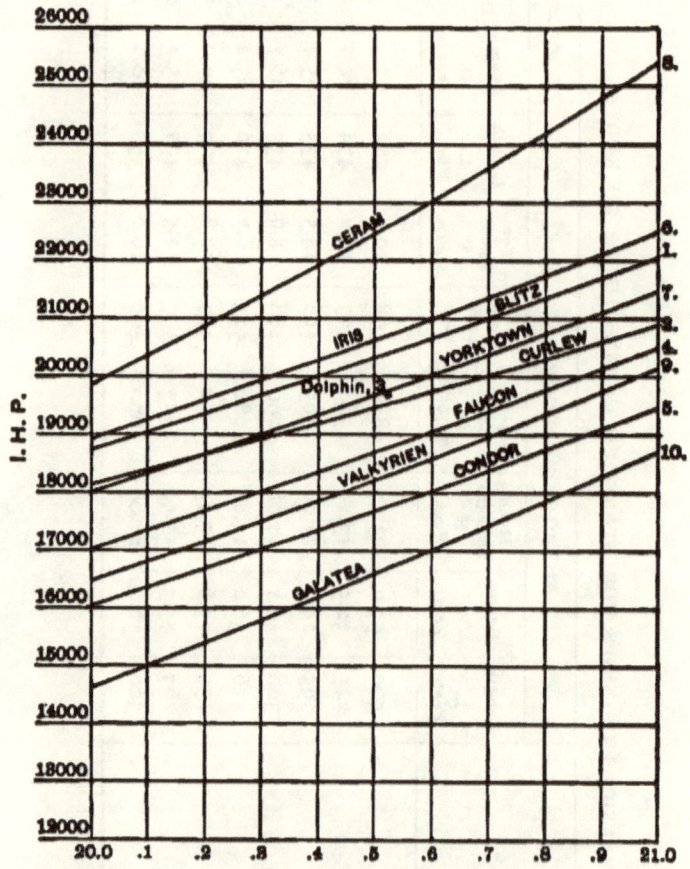

Fig. 57.— STANDARD CURVES. 20 TO 21 KNOTS.

§ 5. STANDARD CURVES OF POWER. 171

DIMENSIONS OF SHIPS FOR STANDARD CURVES 20 TO 21 KNOTS.

Number.	Name.	At Trial Displacement.						At 10,000 Tons.				Top Speed V. Knots.
		Length L. Feet.	Beam B. Feet.	Mean Draught H. Feet.	Displacement D. Tons.	Block Coefficients.	Top Speed V. Knots.	Length L. Feet.	Beam B. Feet.	Mean Draught H. Feet.		
1	Blitz	245	32.5	13.29	1382	.457	16.2	474	62.9	25.75		22.5
2	Curlew	195	28.0	11.00	785	.458	14.5	455	65.4	20.70		22.2
3	Dolphin	240	32.0	13.75	1520	.504	15.0	450	60.0	25.76		20.5
4	Faucon	223	29.2	13.58	1259	.500	17.0	446	58.4	27.16		24.0
5	Condor	223	29.2	13.58	1259	.500	17.7	446	58.4	27.16		25.0
6	Iris	300	46.1	17.08	3290	.488	18.6	435	66.7	24.74		22.4
7	Yorktown	226	36.0	13.84	1680	.523	16.7	410	65.2	25.08		22.4
8	Ceram	152	25.6	9.45	512	.487	12.8	409	69.0	25.45		21.0
9	Valkyrien	267	43.3	17.50	2923	.507	17.5	402	65.2	26.37		21.5
10	Galatea	300	56.0	21.00	5000	.496	19.0	378	70.6	26.46		21.3

172 RESISTANCE OF SHIPS. § 5.

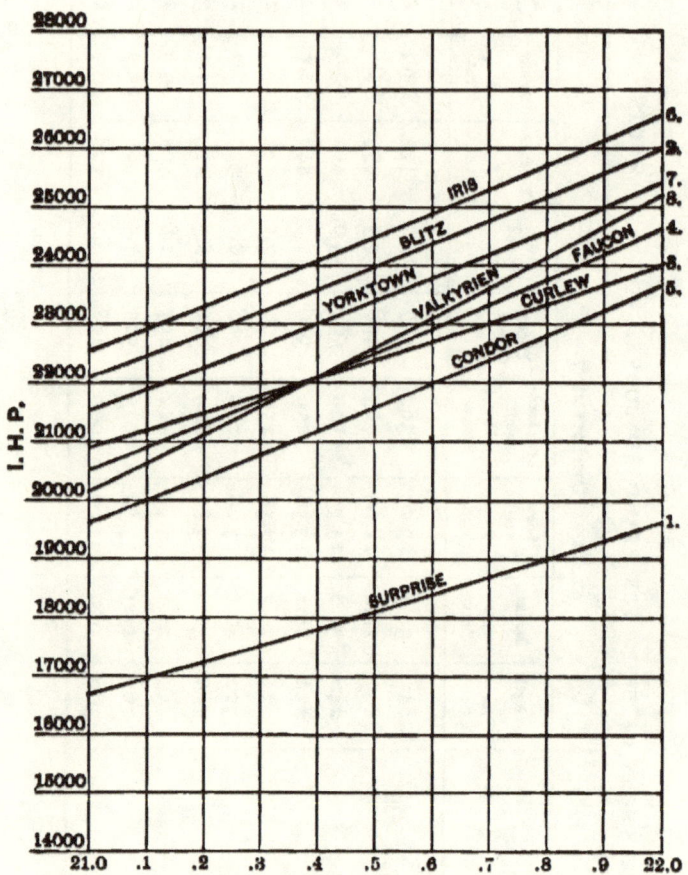

Fig. 58.—STANDARD CURVES. 21 TO 22 KNOTS.

§ 5.   STANDARD CURVES OF POWER.   173

DIMENSIONS OF SHIPS FOR STANDARD CURVES 21 TO 22 KNOTS.

Number	Name	At Trial Displacement.						At 10,000 Tons.			
		Length L. Feet.	Beam B. Feet.	Mean Draught H. Feet.	Displacement D. Tons.	Block Coefficients.	Top Speed V. Knots.	Length L. Feet.	Beam B. Feet.	Mean Draught H. Feet.	Top Speed V. Knots.
1	Surprise	250	32.5	12.29	1306	.458	17.9	493	64.1	24.22	25.1
2	Blitz	245	32.5	13.29	1382	.457	16.2	474	62.9	25.75	22.5
3	Curlew	195	28.0	11.00	785	.458	14.5	455	65.4	20.70	22.2
4	Faucon	223	29.2	13.58	1259	.500	17.0	446	58.4	27.16	24.0
5	Condor	223	29.2	13.58	1259	.500	17.7	446	58.4	27.16	25.0
6	Iris	300	46.1	17.08	3290	.488	18.6	435	66.7	24.74	22.4
7	Yorktown	226	36.0	13.84	1680	.523	16.7	410	65.2	25.08	22.4
8	Valkyrien	267	43.3	17.50	2923	.507	17.5	402	65.2	26.37	21.5

174 RESISTANCE OF SHIPS. § 5.

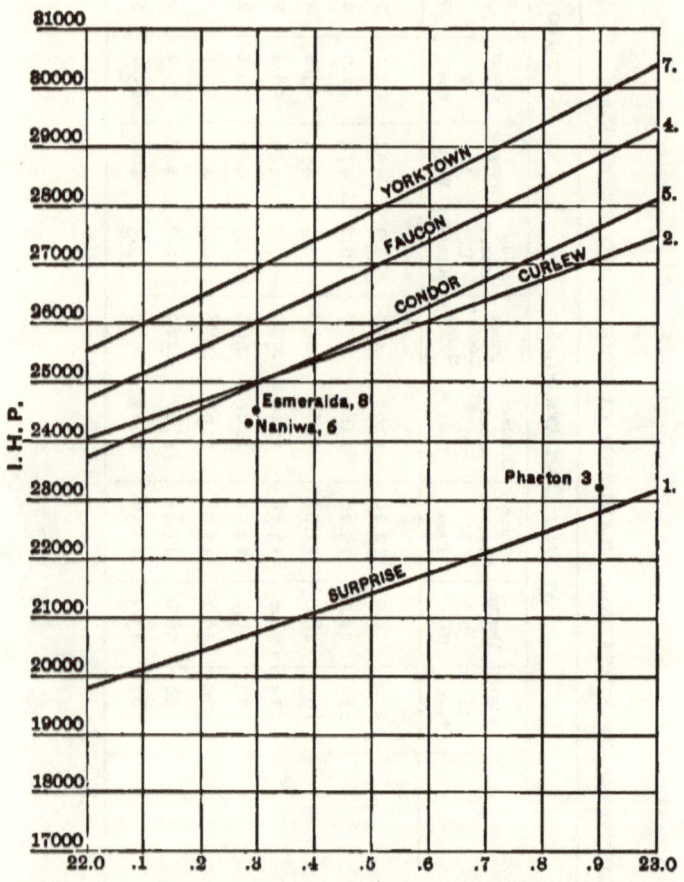

Fig. 59.—STANDARD CURVES. 22 TO 23 KNOTS.

STANDARD CURVES OF POWER. §5. 175

DIMENSIONS OF SHIPS FOR STANDARD CURVES 22 TO 23 KNOTS.

Number	Name	At Trial Displacement.						At 10,000 Tons.			
		Length L. Feet.	Beam B. Feet.	Mean Draught H. Feet.	Displacement D. Tons.	Block Coefficients.	Top Speed V. Knots.	Length L. Feet.	Beam B. Feet.	Mean Draught H. Feet.	Top Speed V. Knots.
1	Surprise	250	32.5	12.29	1306	.458	17.9	493	64.1	24.22	25.1
2	Curlew	195	28.0	11.00	785	.458	14.5	455	65.4	20.70	22.2
3	Phaëton	300	46.0	16.00	2952	.468	18.7	451	69.1	24.03	22.9
4	Faucon	223	29.2	13.58	1259	.500	17.0	446	58.4	27.16	24.0
5	Condor	223	29.2	13.58	1259	.500	17.7	446	58.4	27.16	25.0
6	Naniwa	300	46.0	18.50	3722	.510	18.9	417	64.0	25.72	22.3
7	Yorktown	226	36.0	13.84	1680	.523	16.7	410	65.2	25.08	22.4
8	Esmeralda	270	42.0	18.50	3000	.501	18.3	403	62.7	27.70	22.3

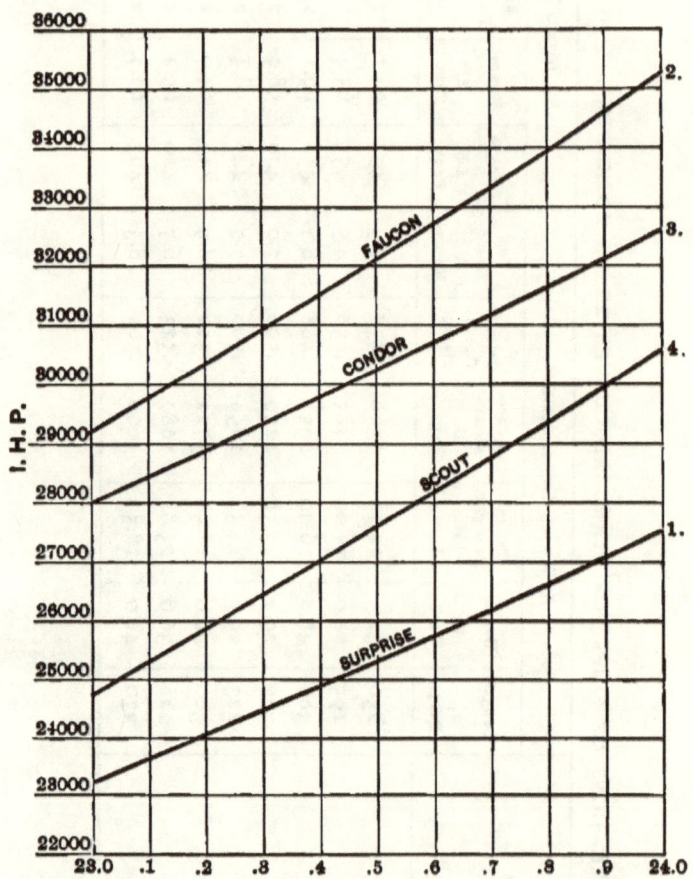

Fig. 60.— STANDARD CURVES. 23 TO 24 KNOTS.

STANDARD CURVES OF POWER.

DIMENSIONS OF SHIPS FOR STANDARD CURVES 23 TO 24 KNOTS.

Number	Name	At Trial Displacement.						At 10,000 Tons.			
		Length L. Feet.	Beam B. Feet.	Mean Draught H. Feet.	Displacement D. Tons.	Block Coefficients.	Top Speed V. Knots.	Length L. Feet.	Beam B. Feet.	Mean Draught H. Feet.	Top Speed V. Knots.
1	Surprise .	250	32.5	13.29	1306	.458	17.9	493	64.1	24.22	25.1
2	Faucon	223	29.2	13.58	1259	.500	17.0	446	58.4	27.16	24.0
3	Condor	223	29.2	13.58	1259	.500	17.7	446	58.4	27.16	25.0
4	Scout .	220	34.0	12.20	1230	.461	17.6	442	68.4	24.50	25.0

178 RESISTANCE OF SHIPS. § 5.

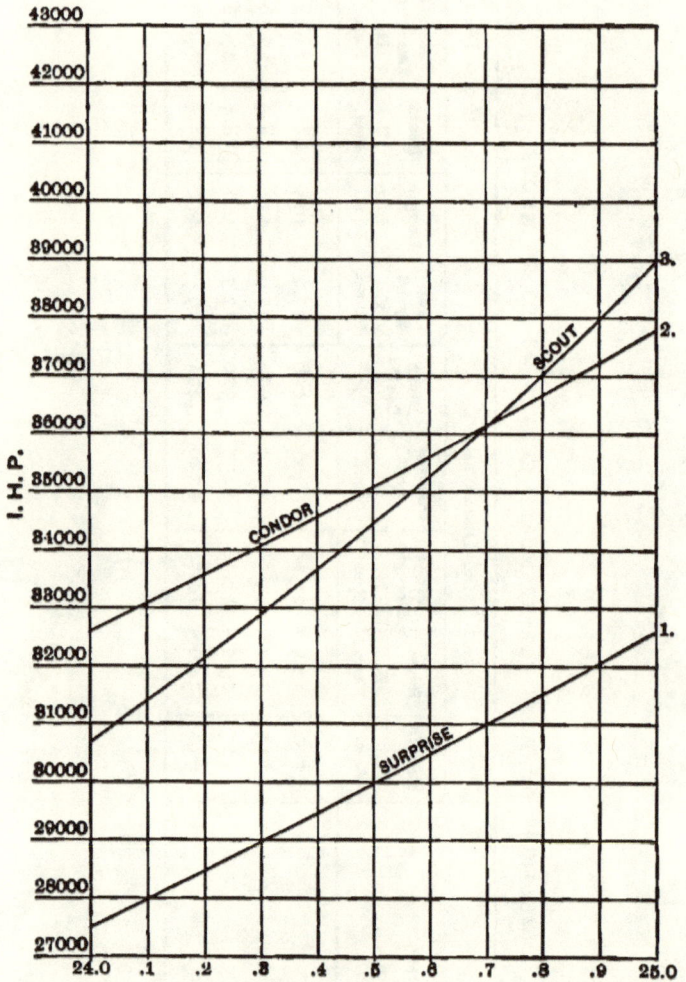

Fig. 61.— STANDARD CURVES. 24 TO 25 KNOTS.

STANDARD CURVES OF POWER.

DIMENSIONS OF SHIPS FOR STANDARD CURVES 24 TO 25 KNOTS.

NUMBER	NAME	At Trial Displacement.						At 10,000 Tons.			
		Length L. Feet.	Beam B. Feet.	Mean Draught H. Feet.	Displacement D. Tons.	Block Coefficients.	Top Speed V. Knots.	Length L. Feet.	Beam B. Feet.	Mean Draught H. Feet.	Top Speed V. Knots.
1	Surprise	250	32.5	12.29	1306	.458	17.9	493	64.1	24.22	25.1
2	Condor	223	29.2	13.58	1259	.500	17.7	446	58.4	27.16	25.0
3	Scout	220	34.0	12.20	1230	.461	17.6	442	68.4	24.50	25.0

§ 6. *Model Tank Method.*

Methods of the second class, in which the effective horse-power and efficiency of propulsion are dealt with separately, while more natural and logical than those of the first class, are somewhat more difficult to apply.

The most accurate method is undoubtedly the model tank method.

The method of determining resistance in the model tank has already been outlined.

Factors to be Determined.
By trying models of the propellers behind the ship, we can determine very closely (in addition to the effective horse-power) the wake factor, the thrust deduction, and the propeller efficiency to be anticipated.

Details of the method of conducting and analysing the model experiments will be found in papers read before the Institution of Naval Architects by the Froudes. Supposing that the values deduced for the wake factor, thrust deduction factor, and propeller efficiency are denoted by w, t, and e_2, we shall have the apparent propeller efficiency $e_1 = \dfrac{1-t}{1-w} e_2$, whence the propeller power $P = \dfrac{E}{e_1} = \dfrac{E(1-w)}{e_2(1-t)}$.

The engine efficiency, or ratio between P and I, being estimated from trials of similar machinery, the proper indicated horse-power is readily determined.

It is evident that the model tank method requires somewhat more complete information than is usually available at the stage of a design where it is necessary to estimate the required I. H. P.

Method not always Applicable.
So that even if model tanks were more numerous, it appears that the usefulness of the method is limited. Unless we are prepared to move very slowly in the early stages of a design, a new model tank can be used only to check an estimate of the I. H. P., already made by some other method.

§ 7. THE INDEPENDENT ESTIMATE METHOD.

But if a model tank has been in existence and active use for some time, there should be recorded at it sufficient data to enable the effective horse-power, wake factor, etc., of any new design of a normal type to be estimated very closely without new experiments.

Unfortunately there are only four or five model tanks in existence, and the results of their experiments are not divulged.

§ 7. *The Independent Estimate Method.*

It is seldom necessary to determine the necessary indicated horse-power of a new design with great accuracy.

So useful results for practical cases may be obtained by what I may call the Independent Estimate Method.

First the effective horse-power is estimated. The skin resistance power is readily calculated from Tables V. and XI., if we know the wetted surface. *Skin Resistance Power.*

This may be obtained very closely by Kirk's analysis. *Wetted Surface.*

A fair approximation to the wetted surface of ships of ordinary form can also be obtained by the formula

$$S = 15.5\sqrt{DL},$$

D being the displacement in tons, and L the length in feet, as usual.

The wave resistance power is readily estimated from a suitable value of the wave resistance constant b_0. *Wave Resistance Power.*

It is best to use values of b_0 obtained from analysis of careful trials of ships, similar to the design in hand. If such are not available, a fair working value of b_0 may be chosen from the information given in § 5, Chapter IV.

Having once determined the effective horse-power, we need to know, in order to determine the values of the I. H. P., the efficiency of propulsion, depending upon the hull efficiency e_3, the propeller efficiency e_2, and the engine efficiency e_1. *Method of Estimating Efficiency of Propulsion.*

These are best determined from analysis of trials of similar ships. In the absence of sufficient information thus obtained, suitable approximate values may be adopted as indicated in Chapter IV. for the engine efficiency.

The hull efficiency may be taken equal to unity.

The propeller efficiency in actual ships often falls below .6. An efficiency of .7 is, however, possible, and if care is bestowed upon the design of the propeller, we should always obtain an efficiency of .65.

Then we have $e = .65 \times 1 \times e_1$.

Since e_1 varies, as we have seen, from .80 to .89, the value of e to be anticipated in designing will vary from .52 to .58, nearly.

If the design contains abnormal features, they should be allowed for in making estimates of the efficiency to be expected.

§ 8. *Comparison of Methods.*

Common Features of All Methods. It will have been made evident by what has gone before that any method for estimating indicated horse-power must be based upon the results of trials of actual ships. Any one of the methods which I have described will give good results in skilful hands, provided it is used in connection with a sufficient amount of data regarding actual trials of ships of diversified types. But the methods described are not all of the same intrinsic value and reliability. The weak points of each method were pointed out after describing it.

Admiralty Coefficient Method. Summing up the subject, it may be said that the admiralty coefficient method is the least reliable of those described.

It is based upon erroneous assumptions, and is essentially untrustworthy.

Kirk's Analysis. Kirk's analysis has all the faults of the admiralty coefficient method except one. While of value for a special purpose,— the determination of the power necessary for the 10-knot cargo steamer,— its range is too limited to allow it to be used with confidence for other types of vessels.

The model tank method, though the most accurate of all those described, is not available for every-day use. *Model Tank Method.*

The extended law of comparison and the independent estimate method are both sufficiently accurate in theory and practice for practical purposes. They must, of course, be used understandingly; but it does not seem possible to devise a method which will give reliable results without a certain amount of intelligence and judgment on the part of the person using it. *Extended Law and Independent Estimate Method.*

The independent estimate method requires more skill and judgment than the extended law. Given the skill, however, it appears to be an essentially more reliable method, especially for low-speed vessels.

§ 9. *Effect of Rough Weather and Foulness.*

Since any method of estimating the I. H. P. required by a given ship for a given speed is based upon trial results, and since trials are almost invariably made in smooth water, and with the ship's bottom clean, the power for a given speed is naturally estimated under these conditions. It should not be forgotten, however, that ordinary seagoing conditions differ from trial conditions. *Seagoing and Trial Conditions Differ.*

It is found, and very naturally too, that more power is developed by a ship during short measured mile trials than can be maintained for a long voyage.

But apart from this, with a given development of power, a ship under ordinary seagoing conditions will not show the same speed as on measured mile trials — made in smooth water. Supposing her bottom clean, in average moderate weather at sea a ship will do from $\frac{1}{4}$ to $\frac{3}{4}$ of a knot less than in dead smooth water, at the same displacement, and for the same power. *Effect of Rough Water.*

In rough and stormy weather the speed of a small, low freeboard vessel may be reduced to almost nothing. But, providing her propellers do not race, necessitating reduced power, it must be stormy weather indeed when the lofty-sided Atlantic "greyhound" loses more than a knot and a half, or thereabouts, through stress of weather.

Absolute size is a great advantage in maintaining speed through rough water, and in the Atlantic liners large displacement is combined with a small proportion of beam to length, the latter being also a feature conducive to the maintenance of speed in a seaway.

Effect of Fouling. The difference between speed on the measured mile and the speed shown in service for the same power is principally due to foulness of bottom.

The exact effect of foulness upon the coefficient of skin friction is a somewhat obscure matter, but marked reduction in speed due to fouling is a matter of common experience. It appears safe to conclude that an amount of fouling is frequently encountered in practice which doubles the coefficient of skin friction, while ships do at times become so covered with barnacles, grass, etc., that the skin friction is four or five times its value for a clean bottom.

Figure 62 shows the effect upon the *Yorktown* power curve of doubling the skin resistance power, making the justifiable assumption that the efficiency of propulsion at a given speed is not appreciably affected by the increased resistance.

It is seen from Figure 62 that in this case the effect of the amount of fouling supposed is to cause always a reduction of two knots, more or less, in the speed shown for a given power.

Indirect Effect of Fouling. It should be pointed out that foulness has an indirect effect upon the speed as well as a direct one. Owing to the increased resistance, it is impossible, with a given effective pressure, to obtain the same number of revolutions with a foul bottom as when the bottom is clean and the resistance less.

It follows that unless the adjustment of cut-off, etc., when working at full power with a clean bottom are such that the

§ 9. EFFECT OF ROUGH WEATHER. 185

mean effective pressure can be increased when fouling ensues, the revolutions and I. H. P. will fall off when the ship becomes foul, and there will be a double loss of speed.

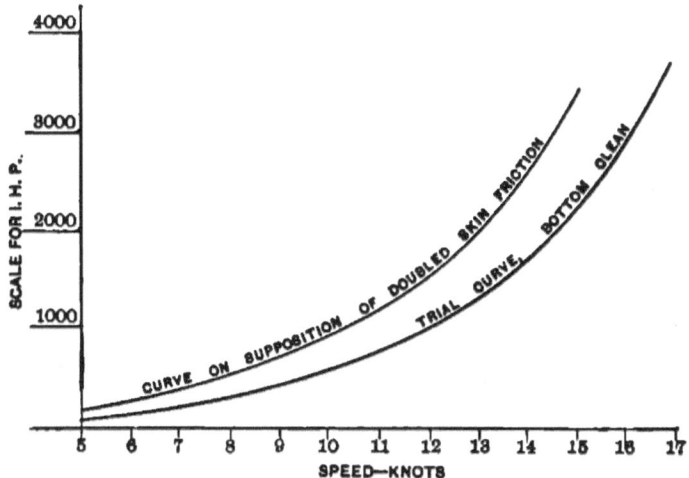

Fig. 62.—CURVES OF I. H. P. FOR YORKTOWN, WITH CLEAN AND WITH FOUL BOTTOM.

There would be some much-needed light thrown upon this question of fouling by a few progressive trials, when foul, of ships whose trial results when clean were known.

Need of Progressive Trials of Foul Ships.

By analysing the trials, the effect of the fouling upon the wake and the slip of the propeller could be determined, as well as the amount of increase in the skin resistance.

CHAPTER VI.

PROPELLER DESIGN.

§ 1. *Influence of Shape of Section and Variation of Pitch.*

Before taking up in detail the question of the design of a propeller to suit given conditions, I shall discuss a question of the greatest practical importance.

Sections of Actual Blade.   A propeller blade, like many other things in this world, has two sides, a fact which should never be lost sight of.

Figure 63 shows six sections of an actual propeller blade whose expanded outline is shown. The blade is somewhat narrow, and being cast-iron, is rather thick. The diameter of the propeller was 14 feet, and the pitch 19 feet. The backs of the sections are circular arcs — a very common shape, especially in merchant work. The sections are taken at radii corresponding to diameter ratios of .2, .3, and so on up to .7.

Lines of Advance.   From Table XIV. I have taken the slip angles for each section corresponding to a slip of 20 per cent, and drawn the corresponding "lines of advance" which show the directions in which the sections advance into still water with the above slip.

Owing to the fact that through interference of the blades the water has a certain sternward velocity when the edge cuts into it, the actual slip angles would be somewhat smaller than shown in the figure.

It is seen that in every section of Figure 63 the back of the blade meets the line of advance at a much greater angle than the front of the blade.

§ 1. INFLUENCE OF SHAPE OF SECTION, ETC. 187

It necessarily follows that the backward pressure on the leading part of the back of the blade is greater than the forward pressure on the leading part of the face of the blade. **Backward Pressure or Thrust.**

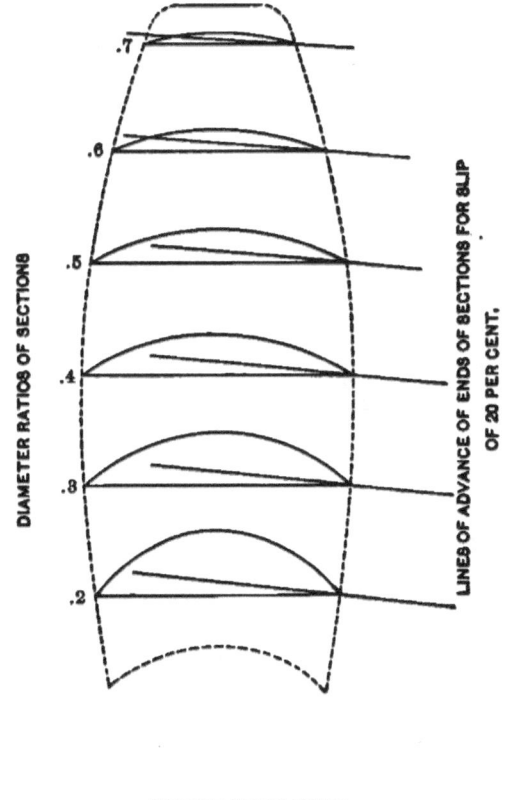

Fig. 63. — SCALE $1'' = 1'8''$.

Some distance from the leading edge the water at the back of the blade will break into eddies, and from that point to the following edge the back of the blade will exert suction, and so help to increase the thrust.

But the actual effective thrust will be the difference between the gross thrust on the blade and the backward thrust on the leading part of the back. Evidently this backward thrust is a source of double loss.

A good deal of power is required to generate it, and once called into existence it neutralises an equal amount of the forward thrust.

Variation of f and a. Referring now to Figure 64, Section A shows on a larger scale the section of Figure 63 at the diameter ratio of .6.

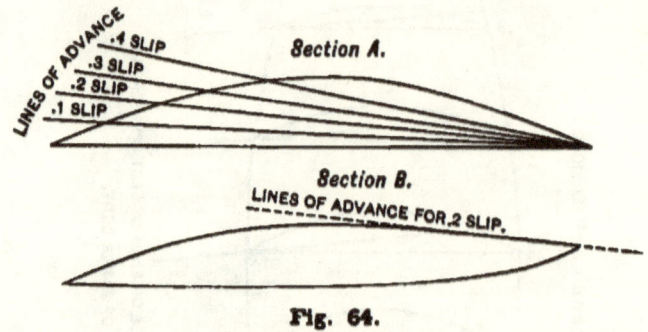

Fig. 64.

On Section A are drawn lines of advance, corresponding to slips of 10, 20, 30, and 40 per cent.

The following conclusions appear obvious:

1. The backward thrust will decrease with increase of slip, which is as much as to say that the thrust coefficient a will increase with increase of slip.

2. As the slip increases, there will be more eddying and less friction on the back of the blade.

3. As the slip increases, the angle at which the face of the blade meets the water will increase, involving diminished friction on the face. (See page 16.)

Then the ratio $\dfrac{f}{a}$ is not a constant quantity, but falls off with increase of slip.

The elder Froude noted the diminution of friction with increase of slip, and stated in 1878 that at a slip of 30 per cent the friction at the back of the blade appeared to be diminished by half.

§ 1.   INFLUENCE OF SHAPE OF SECTION, ETC.   189

It will be necessary to consider the result of the variation of $\frac{f}{a}$, but before so doing I wish to call attention to Section B of Figure 64.

Section B is simply Section A distorted in such a manner that the leading portion of the back of the section is a straight line tangent to the line of advance for 20 per cent slip. *Improved Section.*

The result is increased pitch of the leading portion of the face, and transfer of the backward thrust to the side of useful thrust. This gain far outweighs the slight loss due to the fact that the leading portion of the face will work with greater slip than before. We have seen that the slip must increase a good deal before the efficiency is seriously diminished.

Owing to the short length and great thickness of Section A, the efficiency of the distorted Section B, while above that of A, will still be less than it would have been if A had not had such a large angle between the face and the back of the blade at the leading edge.

The advantages of thin blades, and of leading edges so thin that the line of advance for a fair working slip (which I take as 20 per cent) will clear the profile of the back with little or no increase of pitch along the leading edge of the face, are obvious.

With such blades the value of a will be greater than with the ordinary shape of section, and the change in a as the slip increases will be slight, involving less variation in $\frac{f}{a}$.

To fix our ideas, when considering the effect of the variation of $\frac{f}{a}$, let us take the case of the *Yorktown* propeller. *Effect of Variation of f and a upon Efficiency.*

Here we took $f = .045$, and deduced $a = 7.8$ for slips in the neighbourhood of 20 per cent.

Then, $\qquad \frac{a}{f} = 184, \quad \frac{f}{a} = .00577.$

For efficiency of the propeller we have

$$e = (1-s)\frac{asX_e - fY_e}{asX_e + fZ_e} = (1-s)\frac{Y_e}{Z_e}\frac{\frac{sX_e}{Y_e} - \frac{f}{a}}{\frac{sX_e}{Z_e} + \frac{f}{a}}.$$

For the *Yorktown* propeller

$$X_e = .0954, \quad Y_e = .1439, \quad Z = .4469.$$

Whence, $\quad e = .322(1-s)\dfrac{.663\,s - \dfrac{f}{a}}{.213\,s - \dfrac{f}{a}}.$

Taking $\dfrac{f}{a}$ constant and equal to .00577, we have the curve of efficiency for $\dfrac{f}{a}$ constant, shown in Figure 65.

Suppose, now, $\dfrac{f}{a}$ is variable, changing steadily from .00750 at zero slip to .00350 at 30 per cent slip.

The resulting curve of efficiency is shown in Figure 65, marked for $\dfrac{f}{a}$ variable.

The curve of efficiency of one of Froude's model propellers is also shown in Figure 65.

Practical Conclusions.

What are the conclusions to be drawn from Figure 65?

1. The curve of efficiency with $\dfrac{f}{a}$ variable resembles the curve of the model propeller more closely than when $\dfrac{f}{a}$ is taken as constant.

2. The maximum efficiency corresponds to a greater slip for $\dfrac{f}{a}$ variable.

3. The maximum efficiency is not much changed, and in the vicinity of practical working slips, say from 15 per cent to 25 per cent, the assumption that $\dfrac{f}{a}$ is constant is sufficiently near the truth for practical purposes.

§ 1. INFLUENCE OF SHAPE OF SECTION, ETC.

There are some facts of experience that accord with the theory set forth above of the influence of the shape of the blade section. *Corroboration of Experience.*

1. It is generally admitted that for the same slip the cargo boat requires an essentially larger propeller than the fast twin-screw vessel, and that the propeller of the cargo boat should work with greater slip than that of the fast twin-screw vessel.

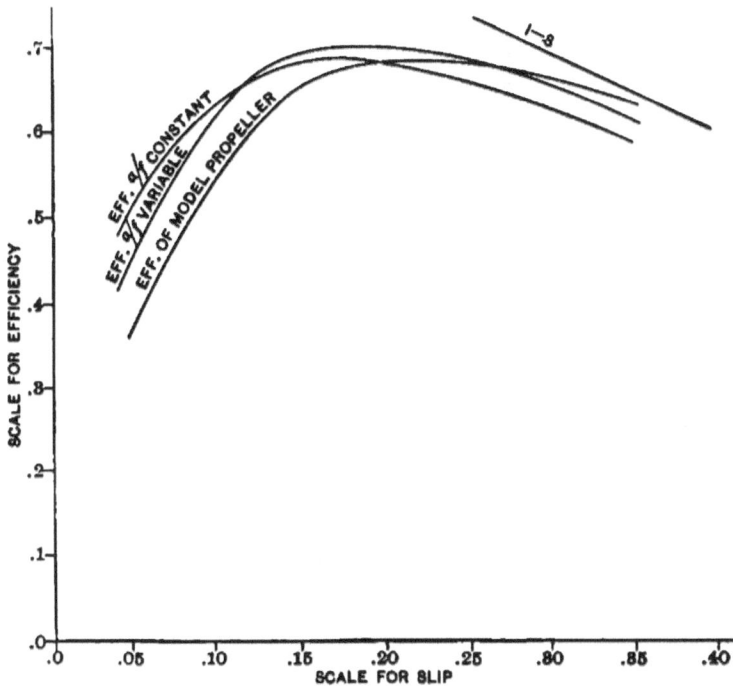

Fig. 65. — EFFICIENCIES ON VARIOUS SUPPOSITIONS.

When I say "larger propeller," I mean, of course, one more than equivalent to the two of the twin-screw vessel.

Now the cargo boat is nearly always fitted with heavy cast-iron propeller blades, while the twin-screw ship will have sharp brass or bronze blades. Hence the cargo boat will have a lower value of a, requiring more blade area to get the thrust with a given slip.

Also the variation in $\frac{f}{a}$ will be greater in the case of the cargo boat and extend further, so that the maximum efficiency will be found at a higher slip.

2. When a cast-iron propeller is replaced by a brass or bronze one, exactly similar except that the blades are thinner and sharper, the engines will not show the same maximum number of revolutions as before. This is because the value of a is increased by the change, more than counterbalancing any slight diminution of f, so that the turning moment necessary to obtain a given number of revolutions is greater than before.

Hence, unless the maximum effective pressure of the engines can be increased, the same maximum revolutions as before cannot be reached.

Axially Increasing Pitch a Source of Loss.

The question of increasing pitch should naturally be considered in connection with the shape of section.

There can be no doubt that axially increasing pitch is a source of dead loss. It is usually adopted with the idea of making the leading portion of the face of the blade tangent, or nearly so, to the line of advance. With a blade of any thickness this involves serious loss, because of the large backward thrust resulting upon the leading portion of the back of the blade.

Even if the blade had no thickness, axially increasing pitch would probably involve loss of efficiency. If $\frac{a}{f}$ were constant over the whole blade, loss of efficiency would certainly ensue.

For, suppose we took two blades A and B of the same size and shape and delivering the same useful work.

Suppose A of uniform pitch, and B of axially increasing pitch.

The slip of every part of A is the same, and can be made the slip of maximum efficiency.

The leading portion of B will have less slip than the

§ 1. INFLUENCE OF SHAPE OF SECTION, ETC.

uniform slip of A, and hence will do its share of work with less efficiency than the corresponding part of A. Part of B about the centre will have the same slip and efficiency as the corresponding part of A. But the rear portion of B must work with greater slip, and hence less efficiency than the corresponding part of A, the total thrust of A and B being the same.

While $\frac{a}{f}$ is probably not constant over the whole blade, it is, without doubt, so nearly constant that the above applies to the case.

Axially increasing pitch is then, without doubt, a source of loss, though the loss is probably not very great. This conclusion is confirmed in a negative way by common experience. Though Rankine and others have recommended increasing pitch, and though it has been tried very largely, and is still much used, it has never prevailed over the uniform pitch, for the simple reason that numerous trials have not shown any appreciable advantage from its use. *Confirmation of Experience.*

While axial increase of pitch is not advantageous, it is probable that a very slight gain would result from a pitch which varied radially in such a manner that each portion of the blade would work at the slip corresponding to its maximum efficiency *Pitch Increasing Radially*

The gain would be so slight, however, that it is not worth while attempting to make it. The only point to be noted in connection with this is that increase of pitch obtained by twisting a blade bodily, as is frequently done with adjustable blades, causes a variation of pitch rather favourable to efficiency than otherwise, while if the pitch be decreased by twisting, there will result a slight tendency toward lessened efficiency.

Hence in designing such propellers we should be careful to estimate the pitch too small rather than too large, in order that change of adjustment, if found necessary, may be made without lessened efficiency.

Table XV. will be found of interest in this connection. It gives the fraction of itself by which the pitch is changed for elements of various diameter ratios from .1 to 1.0, and amounts of twist varying from 1 to 6 degrees.

The pitch of a twisted blade as usually stated is that of the tip.

It would be preferable to take as the true new pitch that at the centre of effort of the blade. The centre of effort may be taken as at about $\frac{2}{3}$ of the radius for ordinary shapes of blade.

Summing up. Summing up what has gone before, it appears that the pitch of the face of a propeller blade should be uniform, unless it is necessary to increase the pitch along the leading edge in order to avoid backward thrust. Even this variation should be avoided, if possible, by making the angle between the face and the back of the blade equal to, or less than, the expected angle of slip. This, however, can seldom be done with blades of cast iron, which are necessarily somewhat thick.

§ 2. *Standard Blade.*

Curved Blades. We frequently find propellers with blades bent backward along a straight line or a curve.

This is with a view to avoid centrifugal action, which is supposed to involve increase of thrust deduction.

The gain can be very slight, and is doubtless nearly or entirely neutralised in practice by increased area, etc., of blade upon the same diameter. So it appears that straight blades will be as good as any other kind, although there are cases when, on account of the nature of the supports or the structure of the ship, the blades of a propeller should be bent backward.

§ 2. STANDARD BLADE.

The final question to be passed upon is that of the shape of blade which should be used. We saw in Chapter II. that the shape of blade was not a very important matter as regards efficiency. *Shape of Blade.*

Now a blade must have a good deal of length at the root in order to secure strength where it merges into the hub, and at the same time the length here, being limited by the size of hub, is usually necessarily less than the greatest length of blade.

Also a blade should have well-rounded ends, with no corners that are sharp or rounded with a small radius.

This is because such outlying and semi-isolated parts of a blade contribute their full share to friction, but not to thrust.

All things considered, a symmetrical blade with moderate curvature of leading and following edges and a well-rounded end appears to be about the best.

Also it is convenient to adopt a blade such that we can express mathematically the curves of its outline.

The standard shape which I propose to use is shown developed in Figure 66.

The maximum length occurs at half-radius. Between this point and the hub the outline is that of an ellipse which, if continued, would touch the axis.

Beyond the half-radius the outline is not elliptical, but represented by a cubic equation of the form $\frac{x^3}{a^3} - \frac{y^3}{b^3} = 1$.

Of course only the part of the curve represented by the above which lies in the first quadrant is used for the blade.

To draw this blade for a given length and radius it is not necessary to have recourse to formulæ. In Figure 66 there is noted the total length of the blade at every twentieth of the radius, the lengths being expressed as fractions of the maximum length at half-radius. *Method of Delineation.*

Hence, having determined the maximum length and the

radius, the developed shape of blade is easily determined by using the fractions of Figure 66.

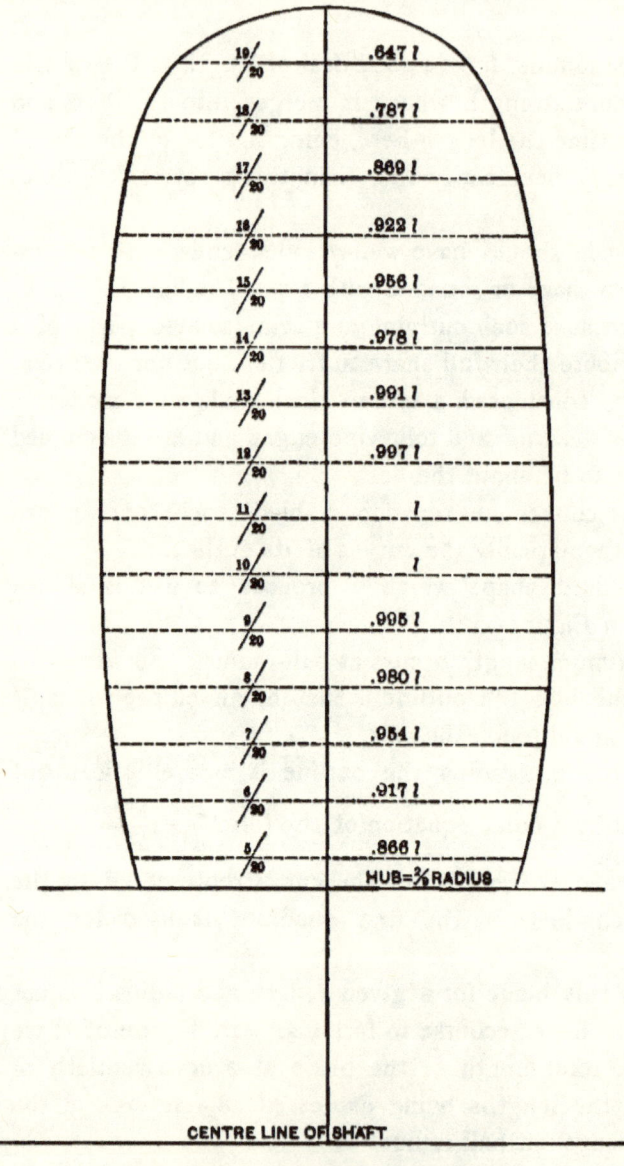

Fig. 66. — STANDARD BLADE.

§ 2. STANDARD BLADE. 197

The diameter of hub is taken as $\frac{2}{9}$ of d, the diameter of the propeller.

The size of hub may be slightly changed, or the standard shape modified near the hub without appreciably changing the characteristics of the blade.

For the standard blade above, if l denote the maximum width or length, we have

$$\text{Area of blade} = .35\, dl.$$

$$\text{Mean width} = .9\, l.$$

$$\text{Whence mean width ratio} = .9\, \frac{l}{d},$$

$\frac{l}{d}$ being the maximum width ratio.

It is, of course, necessary to be able to determine the characteristics of the standard blade without detailed calculations for every case.

We have seen that in order to determine the X characteristic we have

$$X_c = \int X \frac{l}{d} \frac{dr}{d},$$

the integration extending over the blade. Evidently, if we double the width of a blade throughout, we double the characteristic. Then denoting the mean width ratio by h, we can write $X_c = hX_f$, where X_f may be called the "function for X characteristic."

Then

$$hX_f = X_c = \int X \frac{l}{d} \frac{dr}{d},$$

or

$$X_f = \frac{1}{h} \int X \frac{l}{d} \frac{dr}{d}$$

Table XVI. gives values of X_f, Y_f, and Z_f for the standard blade, whence the characteristics for any desired maximum or mean width ratio are readily deduced.

§ 3. Standard Slip.

It is obvious that the problem of propeller design in the ordinary run of cases would be materially simplified if we knew in advance the best slip at which to work. By "best slip" I do not mean the slip which corresponds to the absolute maximum efficiency of the propeller, but the slip which will give the best all-round results.

Assuming $\frac{f}{a}$ to be constant, we have for the efficiency

$$e = (1-s) \frac{asX_c - fY_c}{asX_c + fZ_c}.$$

Since the functions of Table XVI. are all proportional to the characteristics, they may be substituted for them in the above. That is, we may write

$$e = (1-s) \frac{asX_f - fY_f}{asX_f + fZ_f}.$$

Maximum Efficiency. Following the methods of Chapter II. in order to determine the maximum efficiency c_m, and the corresponding slip s_m, and denoting $\frac{a}{f} X_f$ by c, we have

$$c_m = \frac{1}{c} \{ \sqrt{Z_f + c} - \sqrt{Z + Y_f} \}^2,$$

$$s_m = \frac{1}{c} \{ -Z_f - \sqrt{(Z_f + c)(Z_f + Y_f)} \}.$$

While $\quad e = (1-s) \dfrac{cs - Y_f}{cs + Z_f}.$

X_f, Y_f, and Z_f are readily determined from Table XVI.

We have for f the standard value .045, and for a I shall adopt the values obtained in Chapter II. from analysis of the trial results for Froude's experimental propeller. These values are below.

§ 3. STANDARD SLIP. 199

For three-bladed propellers $a = 9.4 - 1.2\,m$.

For four-bladed propellers $a = 8.4 - 1.0\,m$.

Where m is the extreme diameter ratio.

These values are slightly higher than will be found with propellers of the ordinary blade section, but probably not far from the true values for the standard blade section recommended in this chapter.

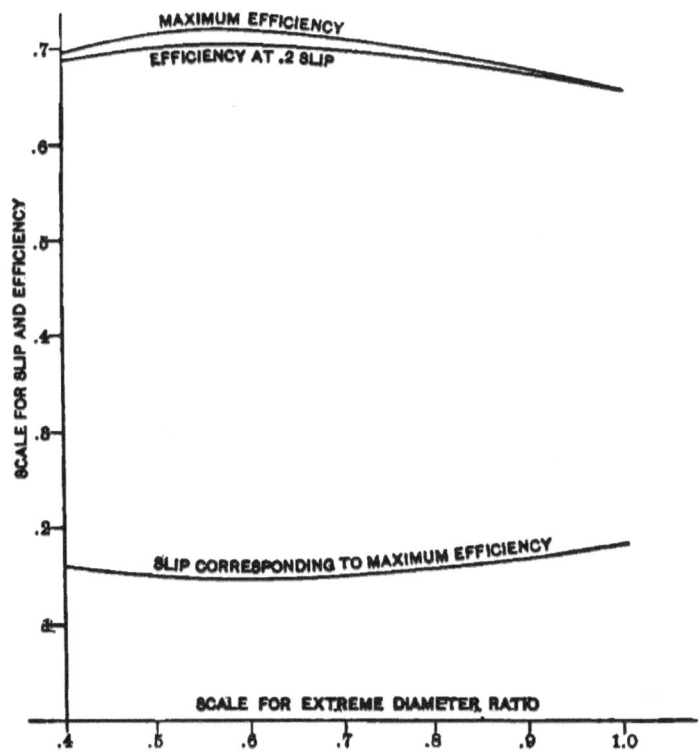

Fig. 67. — EFFICIENCY AND SLIP. THREE-BLADED PROPELLERS.

The calculation of the data necessary to plot Figures 67 and 68 is now easy. Figure 67 refers to three-bladed, and Figure 68 to four-bladed, propellers.

Each figure shows a curve of maximum efficiency and corresponding slip and of efficiency at 20 per cent slip, all plotted upon values of extreme diameter ratio.

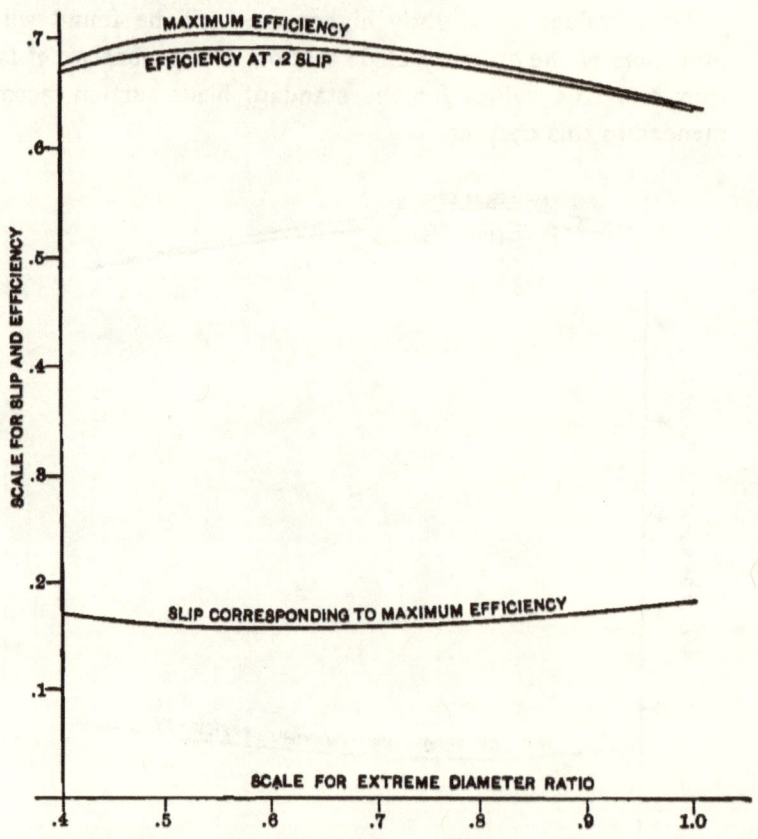

Fig. 68.— EFFICIENCY AND SLIP. FOUR-BLADED PROPELLERS.

These curves are upon the supposition that $\frac{a}{f}$ is constant. We have seen that the effect of the variation in $\frac{a}{f}$ is to increase the slip corresponding to maximum efficiency.

Twenty per cent Slip Standard. All things considered, then, it appears reasonable to aim always at a "standard slip" of 20 per cent in designing. In the best propellers the maximum efficiency will probably be found at a somewhat smaller slip, but the falling off at 20 per cent will be immaterial.

If we were designing propellers for ships which ran always in smooth water with clean bottoms, it would be better to adopt a slightly higher standard slip with a view to securing the greater engine economy which results from quick-running machinery. The final economy as regards coal consumed by the ship in steaming a given distance would not be changed, but the advantages of smaller, lighter, and cheaper engines would be secured. *Considerations Influencing Choice of Slip.*

The propeller, however, must be capable of working at the increased slip found in rough weather, or when the bottom is foul, without a marked loss of efficiency. For this reason a standard slip of 20 per cent seems best for all-round purposes, but the reasons for its adoption should be clearly understood, as a different slip should often be adopted to meet special circumstances.

§ 4. *Design of a Propeller.*

Having disposed of the question of slip, we are now in a position to undertake the design of a propeller to suit a given ship; that is, to settle the diameter, pitch, width ratio, and revolutions, unless the latter are already fixed. *Points to be Settled in Designing a Propeller.*

To fix our ideas, let us suppose that we wish to design a propeller for a 20-knot twin-screw ship of 10,000 indicated horse-power.

Suppose that we have a right to anticipate an engine efficiency of 88 per cent, and estimate by the methods previously given that the wake factor will be 10 per cent.

Then each screw must absorb 4400 horse-power. The speed of ship minus speed of a 10 per cent wake will be 18 knots, or 1824 feet per minute.

The fundamental formula needed is

$$P = 3n\left(\frac{pR}{1000}\right)^3 d^2 (asX_e + fZ_e).$$

The standard slip being .2, we have

$$.8pR = 1824.$$

Whence, $\quad pR = 2280;$

$$\left(\frac{pR}{1000}\right)^3 = 11.852.$$

Then $\quad P = 35.556\, nd^2(.2\, aX_e + fZ_e)$

$$= 7.111\, nd^2(aX_e + 5fZ_e).$$

Revolutions not Fixed. The method of procedure from this point depends upon whether or no the revolutions are fixed. Suppose first that they are not fixed.

In that case we may assume the mean width ratio. Fair maximum working values appear to be .18 for four-bladed propellers, and .2 for three-bladed propellers.

Let us investigate first four-bladed propellers.

Since $X_e = .18\, X_f$, and so on, we have

$$P = 7.111 \times 4 \times d^2 \times .18(aX_f + 5fZ_f)$$

$$= 5.12\, d^2(aX_f + 5fZ_f).$$

Let the extreme diameter ratio $= .5$.

Then $\quad a = 8.4 - 1.0 \times .5 = 7.9.$

$\qquad X_f$ (Table XVI.) $= .236.$

$\qquad f$ standard $\quad = .045.$

$\qquad Z_f$ (Table XVI.) $= .65.$

$\qquad Y_f$ (Table XVI.) $= .557.$

$\qquad P$ (given) $\quad = 4400.$

§ 4. DESIGN OF A PROPELLER. 203

Then $\quad 4400 = 5.12\, d^2(1.864 + .146).$

Whence, $\quad d^2 = \dfrac{4400}{5.12 \times 2.010.} = 427.5,$

$d = 20.68.$

Now, $\quad \dfrac{d}{p} = .5.$

Whence, $\quad p = 2\, d = 41.36,$

and $\quad pR = 2280.$

Whence, $\quad R = \dfrac{2280}{41.36} = 55.12.$

The efficiency from Figure 68 would be about 69 per cent.

Assuming other values of extreme diameter ratio, and hence deducing other corresponding values of d, etc., we obtain data to plot Figure 69, which shows in full lines curves of pitch, revolutions, and efficiency, plotted upon diameter. This is for four-bladed propellers of mean width ratio $=.18$.

If we try three-bladed propellers of mean width ratio $=.20$, we have

$$P = 7.111 \times 3 \times d^2 \times .2(aX_f + 5fZ_f),$$

$$= 4.267\, d^2(aX_f + 5fZ_f).$$

Proceeding as before, we obtain the data for the dotted lines of Figure 69.

From the curves of Figure 69, suitable values of diameter, pitch, and revolutions may be selected.

If, for any reason, it is not possible to adopt suitable values of the above quantities from Figure 69, the circumstances of the case will usually indicate whether we must adopt a slip different from the standard .2, or change the values of mean width ratio used in Figure 69.

Revolutions Fixed in Advance. In practice, the revolutions at which the propeller is to work are usually fixed beforehand, and it is necessary to proportion the propeller to suit.

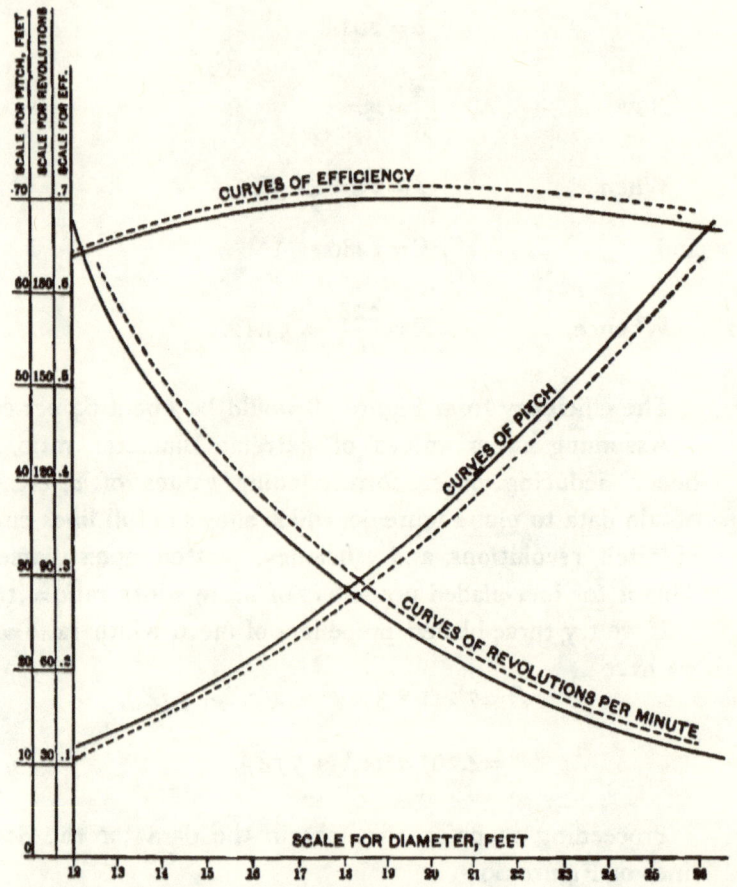

Fig. 69. — FOUR-BLADED PROPELLER, FULL LINES. THREE-BLADED PROPELLER, DOTTED LINES.

Thus, suppose in the preceding, the revolutions had been fixed beforehand at 120 per minute.

Retaining the standard slip of .2, we still have $pR = 2280$; and if $R = 120$, $p = 19$.

§ 4.   DESIGN OF A PROPELLER.

If h denote the mean width ratio, we have

$$P = 4400 = 7.111\, nd^2 h(aX_{\prime} + 5 fZ_{\prime}).$$

$X_{\prime}$ and $Z_{\prime}$ are functions of the diameter ratio, and hence not known in advance.

Assume a diameter ratio of .8.

Then for four-bladed propellers

$$n = 4;$$
$$d = .8 \times 19 = 15.2;$$
$$d^2 = 231, \text{ nearly};$$
$$a = 7.6;$$
$$X_{\prime} = .461;$$
$$Z_{\prime} = 2.04.$$

Everything being now known except h, we find on solving,

$$h = .169.$$

And from Figure 68 the efficiency for .8 diameter ratio = .675.

Proceeding thus for other assumed values of diameter ratio, we obtain the data needed to plot the full curves of Figure 70, which show, plotted upon diameters, the necessary mean width ratios, and resulting efficiencies at every diameter.

The dotted lines refer to three-bladed propellers, and are obtained in precisely the same manner as the full lines.

In choosing a suitable diameter from Figure 70, it is important to remember that the mean width ratio should not exceed .18 for four-bladed propellers, and .20 for three-bladed propellers. While wider blades are often used, there can be no doubt that wide blades greatly aggravate the evil of interference, and that up to a certain point narrow blades are *per se* more efficient. But wide blades mean small diameter, and hence in many cases increased efficiency.

Considerations Influencing Mean Width.

In the case shown in Figure 70 it is desirable, from the point of view of efficiency, to keep the diameter as small as possible.

Values Chosen. Accordingly, adopting a four-bladed propeller and the limiting mean width ratio of .18, we have a diameter corresponding of 14.93 feet, or 14′ 11″, very nearly.

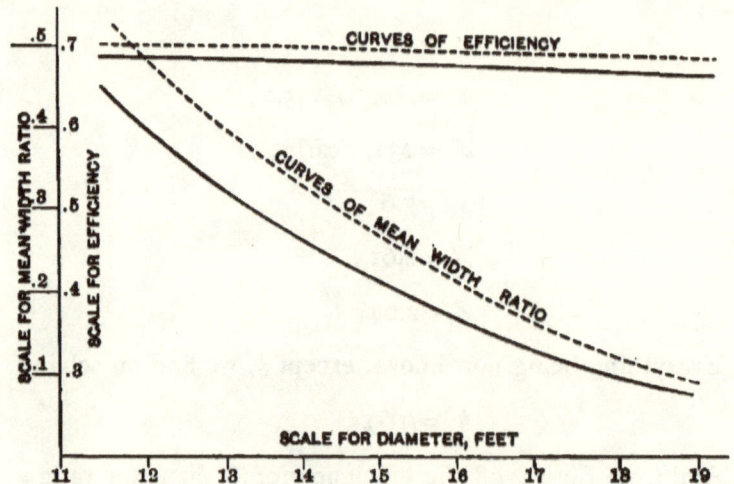

Fig. 70. — CURVES OF EFFICIENCY AND WIDTH RATIO, FULL LINES FOR 4 BLADES, DOTTED FOR 3 BLADES.

The mean width ratio being .18, the mean width = .18 × 14.93 = 2.7, nearly.

The extreme width then $l = \dfrac{2'.7}{.9} = 3'$, and the outline of the blade can be readily plotted from the fractions given in Figure 66.

Coefficients and Constants need not be Exact. It should be remembered that throughout the above we are working with approximate quantities.

Fortunately our approximations may be appreciably in error without seriously impairing the value of the conclusions reached.

For any slip between 15 and 25 per cent the efficiency of a propeller is practically unchanged from the maximum.

§ 4. DESIGN OF A PROPELLER.

The approximate quantities used in the preceding will seldom or never be so much in error that the variation of slip will be more than 2 or 3 per cent above or below the standard 20 per cent.

If it happens that the estimate of wake or horse-power is so much in error that the engine at full power works off the steam too fast, or not fast enough, it is very easy, by adjusting the pitch of the propeller, to hold the revolutions at the proper number. We have seen that this twisting of the blades will not materially affect the efficiency.

It is well known that the propellers of cargo steamers usually work with rather a high slip. *Comparison of Propellers as above and those now fitted.*

It would seem, then, that the adoption of a standard slip of .2 would result in larger propellers than are now generally fitted.

But if the section recommended in the early part of this chapter be used, it will be found that the sizes of propeller will be but little, if any, increased.

This is because with the segmental section commonly used the value of a is often reduced to six, or thereabouts, while with the section I propose the values of a deduced from Froude's experiments will be found very close to the truth.

It will be well to tabulate these values of a for three and four bladed propellers, and the efficiencies at the standard slip of 20 per cent.

They are given below for the range of diameter ratio likely to occur in practice.

In design work the above values of a are only to be used when reliable constants from trials of propellers, similar to the one being designed, are not available.

There is some reason to believe that, owing to minor influences which I have not considered, the maximum efficiency possible corresponds to a somewhat higher diameter ratio than indicated by the above table, probably about .65 to *Absolute Maximum Efficiency.*

.7. This is a matter of no practical importance, however, since from .5 to .8 the efficiency is practically constant.

It should be noted that this is independent of the values of a and f. Possible changes in a and f would change the efficiency for each diameter ratio, but the variation of efficiency with variation of diameter ratio would not appreciably change.

Extreme Diameter Ratio.	Three Blades.		Four Blades.	
	Value of a.	Efficiency at .2 Slip.	Value of a.	Efficiency at .2 Slip.
.4	8.92	.688	8.00	.676
.5	8.80	.702	7.90	.692
.6	8.68	.704	7.80	.694
.7	8.56	.698	7.70	.687
.8	8.44	.686	7.60	.675
.9	8.32	.672	7.50	.660
1.0	8.20	.657	7.40	.643

§ 5. *Strength of Propeller Blades.*

It is, of course, essential that the thickness of a propeller blade should be such that the blade will not break. At the same time it is very desirable that the blade should be no thicker than required for strength.

I propose, then, by the methods already used, to investigate the stress upon a propeller blade, and the strength required.

To do this it is necessary to return to first principles.

Elementary Thrust, Moment, etc. We have seen (page 70) that the elementary thrust upon a small area, dA of a blade, situated at radius r, is given by

$$dT = p^2 R^2 \left[as \frac{\pi^2 y^2 \sqrt{\pi^2 y^2 + (1-s)^2}}{1 + \pi^2 y^2} - f\sqrt{1 + \pi^2 y^2} d \right] dA,$$

§ 5.   STRENGTH OF PROPELLER BLADES.   209

or using the notation subsequently adopted,

$$dT = p^2R^2(asX - fY)ldr,$$

where l is the width of the blade at radius r.

The moment of this elementary thrust about a line through the axis, parallel to the middle of the element $= dT \times r$, very nearly.

Now $r = \frac{1}{2}py$, where y is the diameter ratio at radius r.

Then moment of thrust $= dT \times r$

$$= \frac{p^3R^2}{2}(asXy - fYy)ldr.$$

The elementary turning force dM (page 70)

$$= p^2R^2\left(as\frac{\pi y\sqrt{\pi^2y^2 + (1-s)^2}}{1 + \pi y^2} + f\pi y\sqrt{1 + \pi^2y^2}\right)dA$$

$$= p^2R^2\left(as\frac{X}{\pi y} + f\frac{Z}{\pi y}\right)ldr.$$

Its moment about the axis $= dM \times r$

$$= dM \times \frac{py}{2}$$

$$= \frac{p^3R^2}{2\pi}\left(asX + fZ\right)ldr.$$

Writing for ldr, $d^2 \frac{l}{d}\frac{dr}{d}$ as on page 78, we have

$$dT = p^2R^2d^2\left(asX\frac{l}{d}\frac{dr}{d} - fY\frac{l}{d}\frac{dr}{d}\right).$$

Moment of $dT = \frac{p^3R^2d^2}{2}\left(asXy\frac{l}{d}\frac{dr}{d} - fYy\frac{l}{d}\frac{dr}{d}\right),$

$$dM = \frac{p^2R^2d^2}{\pi}\left(as\frac{X}{y}\frac{l}{d}\frac{dr}{d} + f\frac{Z}{y}\frac{l}{d}\frac{dr}{d}\right).$$

Thrust Transverse Force and Moments.

Moment of dM $= \dfrac{p^3 R^2 d^2}{2\pi}\left(asX\dfrac{l}{d}\dfrac{dr}{d} + fZ\dfrac{l}{d}\dfrac{dr}{d}\right).$

Integrating in order to determine the thrust, etc., for the whole blade, we have

Thrust $= p^2 R^2 d^2 \left[as \int X \dfrac{l}{d}\dfrac{dr}{d} - f\int Y \dfrac{l}{d}\dfrac{dr}{d}\right]$

Moment of thrust $= \dfrac{p^3 R^2 d^2}{2}\left[as\int Xy\dfrac{l}{d}\dfrac{dr}{d} - f\int Yy\dfrac{l}{d}\dfrac{dr}{d}\right].$

Transverse force $= \dfrac{p^2 R^2 d^2}{\pi}\left[as\int \dfrac{X}{y}\dfrac{l}{d}\dfrac{dr}{d} + f\int \dfrac{Z}{y}\dfrac{l}{d}\dfrac{dr}{d}\right].$

Transverse moment $= \dfrac{p^3 R^2 d^2}{2\pi}\left[as\int X\dfrac{l}{d}\dfrac{dr}{d} + f\int Z\dfrac{l}{d}\dfrac{dr}{d}\right].$

Now $\int X\dfrac{l}{d}\dfrac{dr}{d}$ we have denoted by X_e, and so for Y_e and Z_e. So let us denote these new quantities in accordance with the same plan; *i.e.*

$\int Xy\dfrac{l}{d}\dfrac{dr}{d}$ by X_y,

$\int \dfrac{X}{y}\dfrac{l}{d}\dfrac{dr}{d}$ by $X_{1/y}$.

$\int Yy\dfrac{l}{d}\dfrac{dr}{d}$ by Y_y.

$\int \dfrac{Z}{y}\dfrac{l}{d}\dfrac{dr}{d}$ by $Z_{1/y}$.

Then thrust $= p^2 R^2 d^2(asX_e - fY_e).$

Moment of thrust $= \dfrac{p^3 R^2 d^2}{2}(asX_y - fY_y).$

Transverse force $= \dfrac{p^2 R^2 d^2}{\pi}(asX_{1/y} + fZ_{1/y}).$

Transverse moment $= \dfrac{p^3 R^2 d^2}{2\pi}(asX_e + fZ_e).$

§ 5. STRENGTH OF PROPELLER BLADES.

Let us call the point at which the thrust must be concentrated to produce the proper moment the centre of thrust, and let its distance from the centre of the shaft be denoted by $k_1 r$, where r is the extreme radius. *Centre of Thrust.*

Then,
$$k_1 r = \frac{\text{moment of thrust}}{\text{thrust}}$$

$$= \frac{p^3 R^2 d^2 \frac{1}{2}}{p^2 R^2 d^2} \left(\frac{as X_t - f Y_t}{as X_e - f Y_e} \right)$$

$$= \frac{p}{2} \left(\frac{as X_t - f Y_t}{as X_e - f Y_e} \right).$$

But $\frac{r}{p} = \frac{1}{2} y_0$, where y_0 is the extreme diameter ratio.

$$k_1 r = \frac{k_1 p y_0}{2} = \frac{p}{2} \left(\frac{as X_t - f Y_t}{as X_e - f Y_e} \right),$$

or
$$k_1 = \frac{1}{y_0} \left(\frac{as X_t - f Y_t}{as X_e - f Y_e} \right).$$

Similarly, regarding the transverse force as concentrated at the centre of transverse force whose distance from the axis is denoted by $k_2 r$, we have *Centre of Transverse Force.*

$$k_2 = \frac{1}{y_0} \frac{as X_e + f Z_e}{as X_{1/t} + f Z_{1/t}}.$$

The functions X_t, $X_{1/t}$, etc., are readily calculated for the standard blade in the same manner as X_e, etc., are calculated.

Their values for a mean width ratio of .18 are given below in Table XVII., together with the characteristics X_e, Y_e, and Z_e for the same mean width ratio of .18.

TABLE XVII.

Values of X_c, $X_{1/y}$, Y_y, $Z_{1/y}$, X_c, Y_c, and Z_c; Standard Blade Mean Width Ratio, .18.

Extreme Diameter Ratio.	X_c	Y_c	Z_c	X_y	$X_{1/y}$	Y_y	$Z_{1/y}$
.4	.031	.095	.087	.009	.115	.023	.222
.5	.043	.100	.117	.015	.128	.032	.312
.6	.056	.108	.169	.023	.143	.042	.415
.7	.069	.118	.251	.033	.156	.052	.525
.8	.083	.128	.367	.046	.161	.062	.647
.9	.092	.140	.516	.061	.173	.081	.797
1.0	.111	.152	.707	.077	.179	.099	.976

Values of k_1 and k_2.

Given the quantities in the above table, and assuming the standard slip of 20 per cent, it is easy to determine the values of k_1 and k_2 for any diameter ratio. The values in question are tabulated below, and shown graphically in Figure 71.

These values are for four-bladed propellers. Those for three-bladed propellers are practically identical.

TABLE XVIII.

Values of k_1 and k_2.

Extreme Diameter Ratio.	k_1.	k_2.	Extreme Diameter Ratio.	k_1.	k_2.
.4	.706	.646	.8	.688	.614
.5	.710	.658	.9	.695	.606
.6	.692	.644	1.0	.696	.600
.7	.684	.625			

§ 5. STRENGTH OF PROPELLER BLADES. 213

This table and Figure 71 afford the information necessary to enable us to determine the positions of the centres of thrust and of transverse force in any case.

Working Formulæ for Thrust and Transverse Force.

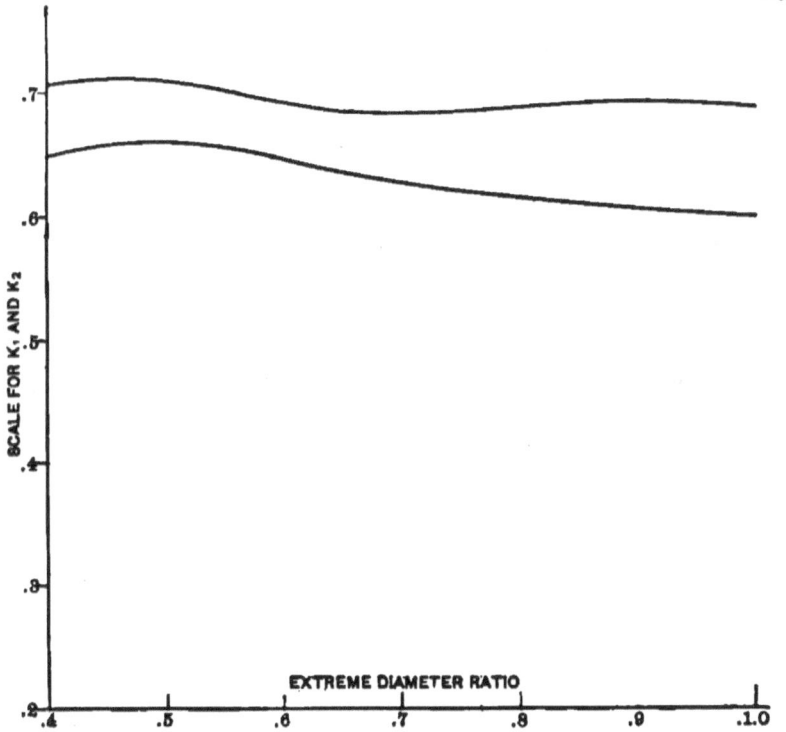

Fig. 71. — VALUES OF K_1 AND K_2.

The values of the transverse force and thrust can be deduced with sufficient approximation from the propeller power P.

Let A denote the transverse force in pounds exerted against a single blade of a propeller of n blades.

Then $A \cdot n \cdot k_2 \cdot \dfrac{d}{2} \cdot 2\pi R = $ work delivered to the propeller in one minute in foot-pounds $= 33{,}000\ P$. Whence

$$A = \frac{33000\,P}{n \cdot k_2 d \pi R}.$$

Let B denote the thrust in pounds upon one blade.

Then $B \cdot n \cdot p(1-s)R$ = useful work done per minute in foot-pounds = $33,000\,U$.

Then
$$B = \frac{33000\,U}{n \cdot p \cdot (1-s)R}.$$

The standard slip being .2, we have $1-s=.8$.

Also it is best in this work to assume a constant efficiency of propeller of .7.

This is slightly above the truth, and allows a slight margin of safety. Making the necessary substitutions, we have

$$B = \frac{33000 \times .7\,P}{.8\,npR} = \frac{28875\,P}{np \cdot R}$$

Application to Practical Case. From the above we can deduce the transverse and longitudinal bending moments upon the root section of the blade, which is the most strained. The *modus operandi* will be best understood from a practical example.

For the four-bladed propeller at 120 revolutions of § 4, we have

$$P = 4400\,;$$
$$p = 19\,;$$
$$d = 14.93\,;$$
$$R = 120\,;$$
$$n = 4\,;$$
$$y_0 = \frac{14.93}{19} = .786.$$

By interpolation in Table XVIII., or directly from Figure 71,

$$k_1 = .687\,; \quad k_1\frac{d}{2} = 5'.13\,;$$

$$k_2 = .615\,; \quad k_2\frac{d}{2} = 4'.59.$$

§ 5. STRENGTH OF PROPELLER BLADES.

From the above we deduce

$$A = 10,485;$$
$$B = 13,930.$$

The most strained section is that at the root. Now the radius of hub $= \tfrac{1}{9} \times 14.93 = 1.66$.

Then transverse moment at root

$$= A\left(k_2 \frac{d}{2} - 1.66\right)$$
$$= 10,485\,(4.59 - 1.66) \times 12$$
$$= 10,485 \times 2.93 \times 12 \text{ in inch-pound units}$$
$$= 368,600.$$

Fore and aft moment at root

$$= B\left(k_1 \frac{d}{2} - 1.66\right) \text{ in feet}$$
$$= 13,930\,(5.13 - 1.66)\,12, \text{ in inch-pound units.}$$
$$13,930 \times 3.47 \times 12 = 580,100.$$

These moments are shown graphically in Figure 72.

By resolving them into two other moments, perpendicular and parallel to the face of the blade at root, we obtain the bending moments upon the root section which are perpendicular and parallel to the face.

These latter moments can also be calculated. Referring to Figure 72,

$$AB = p = 19';$$
$$CB = 2\pi \times 1.66 = 10'.47;$$
$$\tan \theta = \frac{19}{10.47} = 1.815;$$
$$\theta = 61° 09';$$
$$\sin \theta = .8759;$$
$$\cos \theta = .4825;$$

216　　　　　RESISTANCE OF SHIPS.　　　　　§ 5.

MT denotes the transverse moment, and MS the fore-and-aft moment.

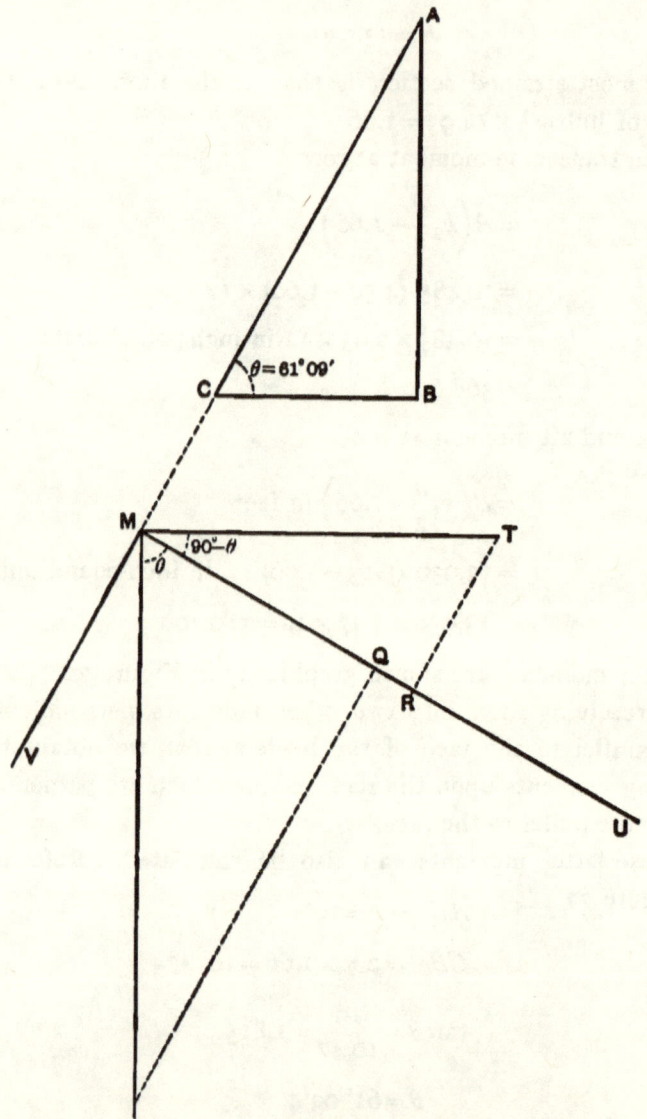

Fig. 72.— BENDING MOMENTS ON ROOT SECTION.

§ 5. STRENGTH OF PROPELLER BLADES.

Draw TR, SQ parallel to the face of the blade to meet a line perpendicular to the face.

Then moment perpendicular to face of blade

$$= MU = MQ + MR$$
$$= MS \cos \theta + MT \cos (90° - \theta)$$
$$= MS \cos \theta + MT \sin \theta$$
$$= 279{,}900 + 322{,}900$$
$$= 602{,}800, \text{ denoted by } M_1.$$

Moment parallel to face of blade

$$= MV = SQ - TR$$
$$= MS \sin \theta - MT \cos \theta$$
$$= 508{,}150 - 177{,}850$$
$$= 330{,}300, \text{ denoted by } M_2.$$

Having arrived at the bending moments upon the root section, it behooves us next to consider the moments of resistance of the latter.

Moments of Inertia, etc., of Root Section.

The root section is usually made a segment of a circle. This is so close to a parabolic segment for the proportions of depth and width found in practice that I shall call it parabolic, since the moments of inertia of a parabolic segment can be expressed very simply.

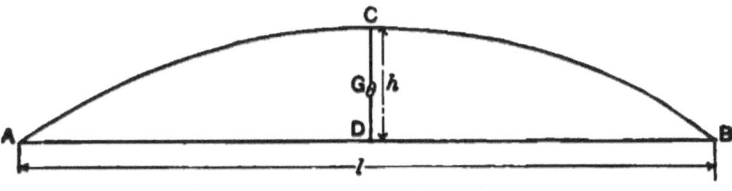

Fig. 73. — PARABOLIC SEGMENT.

Figure 73 shows a parabolic segment of width or length of base AB denoted by l, and of depth or thickness CD denoted by h.

The properties of this segment which we shall need are as follows :
$$\text{Area} = \tfrac{2}{3} AB \times CD = \tfrac{2}{3} lh.$$

DG, or distance of centre of gravity of the section from the face, $= \tfrac{2}{5} CD = \tfrac{2}{5} h$.

Moment of inertia of section about a line through G parallel to the face $= \tfrac{8}{175} lh^3$: denote this by I_1.

Moment of inertia of section about a line through G perpendicular to the face $= \tfrac{1}{30} l^3 h$: denote this by I_2.

Now M_1 denotes the bending moment perpendicular to the face of the blade, and M_2 the moment parallel to the face.

Stresses on Section. Then by the well-known formulæ of applied mechanics for the strength of beams, we have

From M_1,

$$\text{Tension at } A \text{ and } B = \tfrac{2}{5} h \times \frac{M_1}{I_1}$$

$$= M_1 \times \frac{\tfrac{2}{5} h}{\tfrac{8}{175} lh^3} = \frac{35}{4} \frac{M_1}{lh^2}.$$

$$\text{Compression at } C = \tfrac{3}{5} h \times \frac{M_1}{I_1} = \frac{105}{8} \frac{M_1}{lh^2}.$$

From M_2,

$$\text{Tension at } A = \tfrac{1}{2} \frac{lM_2}{I_2} = \frac{15\, M_2}{l^2 h}.$$

$$\text{Compression at } B = \tfrac{1}{2} \frac{lM_2}{I_2} = -\frac{15\, M_2}{l^2 h}.$$

In all practical cases the maximum tension, which I shall denote by f_1, is at A, and the maximum compression, denoted by f_2, at C.

Then,
$$f_1 = \frac{35}{4} \frac{M_1}{lh^2} + \frac{15\, M_2}{l^2 h},$$

$$f_2 = \frac{105}{8} \frac{M_1}{lh^2}.$$

l is fixed by the width of the propeller near the boss, so we can from the above determine the requisite thickness for a given maximum tension or compression.

§ 5. STRENGTH OF PROPELLER BLADES.

The amount of these maximum strains allowable depends upon the material. An ample margin of safety should be allowed in all cases, and the fact that these are fluctuating stresses borne in mind. *(Maximum Stresses Allowable.)*

All things considered, the following seem allowable working stresses:

TABLE XIX.

Material.	Working Stress Allowed. Pounds per Sq. In.	
	In Tension.	In Compression.
Cast iron	2000	6000
Cast steel	5000	10000
Composition	3000	4000
Manganese and phosphor bronze	5000	6000

Let us determine the root thickness necessary for the propeller discussed in the preceding part of this section, supposing cast iron the material used. *(Thickness at Root for Various Materials.)*

We have
$M_1 = 602,800$;
$M_2 = 330,300$;
$f_1 = 2000$;
$f_2 = 6000$.

The maximum width of blade is 3 feet, or 36 inches. Then the width at root (Fig. 66)

$$= 36 \times .86 = 31''.$$

From the equation for maximum tension we have

$$2000 = \frac{35}{4} \times \frac{602800}{31 h^2} + \frac{15 \times 330300}{961 h},$$

or, $h^2 - 2.578 h = 85.09$;

whence, $h = 10''.5$, about.

From the equation for maximum compression,

$$6000 = \frac{105}{8} \times \frac{602800}{31\,h^2};$$

$$h^2 = 43.42;$$

whence, $h = 6''.6.$

Of course, the greatest value of h must be used, since the blade must be strong enough to stand both tension and compression.

I have calculated the values of h for the above propeller blade for the various materials and stresses of Table XIX. The results are of interest, and are given below in Table XX.

TABLE XX.

Root Thickness of Blades of Various Materials.

Material of Blades.	Thickness of Root from Consideration of	
	Tension.	Compression.
Cast iron	$10''.5$	$6''.6$
Cast steel	$6''.4$	$5''.1$
Composition	$8''.4$	$8''.1$
Manganese or phosphor bronze	$6''.4$	$6''.6$

Having settled the thickness at root, the thickness near the tip is made just enough to allow the blade to be cast easily, and the back of the blade is given a uniform taper.

By this method, if the root is thick enough, the thickness will be sufficient throughout.

The formulæ given above for the segmental section apply with ample approximation to the modified section, shaped to avoid backward thrust, which I have recommended.

The extra thickness of blade necessitated when cast iron is

§ 5. STRENGTH OF PROPELLER BLADES.

used is very objectionable. Manganese and phosphor bronze are usually used for propellers at present in high-class work. It appears probable that in time cast iron will cease to be used except for special cases.

For a tug, or ship exposed frequently to danger of damage to its propeller, cast iron is preferable. Its brittleness now becomes a virtue. When a cast-iron propeller strikes an obstruction there is seldom more damage done than is involved in the loss of a piece or the whole of a blade. If a propeller of stronger and tougher material strikes an obstruction, there is danger of serious injury to the shaft or engines.

Table III.

Froude's Frictional Constants for Salt-Water, Paraffin, or Smoothly-Painted Surfaces.

Length of Vessel or Model in Feet.	Coefficient of Friction. f.	Power according to which Friction varies. m.	Length of Vessel or Model in Feet.	Coefficient of Friction. f.	Power according to which Friction varies. m.
8	.01197	1.825	80	.00933	1.825
9	.01177		90	.00928	
10	.01161		100	.00923	
12	.01131		120	.00916	
14	.01106		140	.00911	
16	.01086		160	.00907	
18	.01069		180	.00904	
20	.01055		200	.00902	
25	.01029		250	.00897	
30	.01010		300	.00892	
35	.00993		350	.00889	
40	.00981		400	.00886	
45	.00971		450	.00883	
50	.00963		500	.00880	
60	.00950		550	.00877	
70	.00940		600	.00874	

Table IV.

Surface Friction Constants for Paraffin Models in Fresh Water.

Length of Model in Feet.	Coefficient of Friction. f.	Power according to which Friction varies. m.	Length of Model in Feet.	Coefficient of Friction. f.	Power according to which Friction varies. m.
2.0	.01176	1.94	12.0	.00908	1.94
3.0	.01123		12.5	.00901	
4.0	.01083		13.0	.00895	
5.0	.01050		13.5	.00889	
6.0	.01022		14.0	.00883	
7.0	.00997		14.5	.00878	
8.0	.00973		15.0	.00873	
9.0	.00953		16.0	.00864	
10.0	.00937		17.0	.00855	
10.5	.00928		18.0	.00847	
11.0	.00920		19.0	.00840	
11.5	.00914		20.0	.00834	

TABLE V.

Surface Friction Constants for Ships in Salt Water of 1.026 Density.

Length of Ship in Feet.	Iron Bottom. Clean and Well Painted.		Copper or Zinc Sheathed.			
			Sheathing Smooth and in Good Condition.		Sheathing Rough and in Bad Condition.	
	f.	m.	f.	m.	f.	m.
10	.01124	1.8530	0.1000	1.9175	.01400	1.8700
20	.01075	1.8490	.00990	1.9000	.01350	1.8610
30	.01018	1.8440	.00903	1.8650	.01310	1.8530
40	.00998	1.8397	.00978	1.8400	.01275	1.8470
50	.00991	1.8357	.00976	1.8300	.01250	1.8430
100	.00970	1.8290	.00966	1.8270	.01200	1.8430
150	.00957	1.8290	.00953	1.8270	.01183	1.8430
200	.00944	1.8290	.00943	1.8270	.01170	1.8430
250	.00933	1.8290	.00936	1.8270	.01160	1.8430
300	.00923	1.8290	.00930	1.8270	.01152	1.8430
350	.00916	1.8290	.00927	1.8270	.01145	1.8430
400	.00910	1.8290	.00926	1.8270	.01140	1.8430
450	.00906	1.8290	.00926	1.8270	.01137	1.8430
500	.00904	1.8290	.00926	1.8270	.01136	1.8430

TABLE VI.

Comparison between Waves in Shallow and Deep Water.

Depth of Water as a Fraction of the Length of Wave.	Ratio between Quantities for Shallow Water and Corresponding Quantities for Deep Water.		
	Length and Velocity for Given Period.	Length for Given Velocity.	Velocity for Given Length.
.01	.063	15.90	.251
.02	.124	8.08	.352
.03	.186	5.376	.431
.04	.246	4.065	.496
.05	.304	3.289	.552
.075	.439	2.277	.663
.10	.557	1.796	.746
.15	.736	1.358	.858
.20	.847	1.180	.920
.25	.917	1.091	.958
.30	.955	1.047	.977
.35	.975	1.026	.987
.40	.987	1.013	.993
.45	.993	1.007	.996
.50	.996	1.004	.998
.55	.998	1.002	.999
.60	.999	1.001	.9999
.75	.9999	1.0001	.99999
1.00	.99999	1.00001	.999999

TABLE VII.
Values of X'.

Diameter Ratio = y. Slip = s.	.1	.2	.3	.4	.5	.6	.7	.8	.9	1.0
.05	.090	.320	.630	.964	1.306	1.648	1.986	2.320	2.656	2.980
.10	.086	.311	.613	.946	1.288	1.630	1.969	2.305	2.638	2.967
.15	.081	.299	.597	.929	1.271	1.614	1.954	2.291	2.625	2.956
.20	.077	.288	.582	.912	1.254	1.598	1.939	2.277	2.612	2.944
.25	.073	.277	.567	.896	1.239	1.583	1.925	2.264	2.600	2.933
.30	.069	.266	.552	.881	1.224	1.569	1.912	2.252	2.589	2.922
.35	.065	.256	.539	.864	1.210	1.556	1.900	2.241	2.579	2.913

Table VIII.

Values of X, Y, and Z.

Diameter Ratio.	X.	Y.	Z.
.1	.077	1.048	.10
2	.288	1.181	.47
.3	.582	1.374	1.22
.4	.912	1.606	2.54
.5	1.254	1.862	4.60
.6	1.598	2.134	7.58
.7	1.939	2.416	11.68
.8	2.277	2.705	17.09
.9	2.612	2.999	23.98
1.0	2.944	3.297	32.54

TABLE IX.

Diameter Ratio.	.2		.4		.6		.8		1.0	
Slip.	X'	X	X'	X	X'	X	X'	X	X'	X
			Comparative Efficiencies obtained by using X' and X as indicated							
.05	.523	.482	.627	.612	.566	.558	.483	.478	.404	.401
.10	.678	.662	.726	.721	.682	.679	.618	.616	.549	.547
.15	.702	.696	.734	.732	.703	.701	.654	.653	.599	.599
.20	.690	.690	.715	.715	.691	.691	.654	.654	.609	.609
.25	.664	.667	.684	.686	.666	.667	.636	.637	.600	.600
.30	.630	.635	.648	.650	.633	.634	.609	.610	.579	.580
.35	.592	.598	.607	.610	.596	.597	.576	.577	.552	.552

1	Speed-knots $= V$	5	6	7
2	Revolutions $= R$	44.8	53.7	62.
3	Indicated horse-power $= I$	96	143	205
4	I_f	43	51	59.
5	$3\text{-}4 \; I\text{-}I_f$	53	92	146
6	$.93 \, (I\text{-}I_f) = P$	49	86	136
7	$(R \div 100)^8$	.090	.155	
8	$as + .211$	1.475	1.502	1.
9	as	1.264	1.291	1.
10	Speed of screw-knots	5.53	6.63	7.
11	Slip for zero wake	.0958	.0950	
12	Add for .10 wake	.0904	.0905	
13	Slip for .10 wake	.1862	.1855	
14	Slip for .20 wake	.2766	.2760	
15	a for zero wake	13.20	13.60	13.
16	a for .10 wake	6.78	6.96	7
17	a for .20 wake	4.56	4.68	4
11$_a$	Apparent slip, s'	.0958	.0950	).
18	Add for .083 wake	.0750	.0751	
19	True slip, s	.171	.170	ad
20	$s - \dfrac{fY_c}{aX_c} = s - .009$	.162	.161	
21	$s + \dfrac{fZ_c}{aX_c} = s + .027$	.198	.197	$\dfrac{9}{7}$
22	Apparent propeller efficiency	.740	.740	
23	Thrust horse-power $= T$	36	64	101
24	$.3129 V^{2.88} = E_s$	30	50	77
25	$.0030707 V^5 = y$	9.6	23.9	51.
26	$E_s \div y = c$	3.125	2.093	1.
27	$T \div y = x$	3.750	2.678	1.
28	Thrust efficiency $= T \div I$	.375	.448	
29	Ratio $c \div e_{16}$	.586	.699	
30	$E_s \div I = e$, for $b = 0$	.313	.350	
31	$.0030707 V^3 \div I =$	.100	.167	
32	e_{16} for $b = 0$	.534	.513	)m
33	Add for $b = .5$	.085	.123	
34	e_{16} for $b = .5$	.619	.636	
35	e_{16} for $b = 0$ faired	.546	.521	
36	e_{16} for $b = .5$ faired	.640	.659	
37	$(1-t)T = E$	33	59	93
24$_a$	E_s	30	50	77
38	$.0030707 V^5 \times b = E_w$	3	9	16
39	b	.312	.377	
40	$e = (1-t)(\text{thrust efficiency})$	.345	.412	

Table XI.

Powers of Speeds needed in Calculations.

Speed V.	$V^{1.85}$	$V^{2.17}$	$V^{2.83}$	V^5	$.0030707\, V^5$
1	1.00	1.0	1.0	1	0.0
2	3.56	4.5	7.1	32	0.1
3	7.47	10.8	22.4	243	0.7
4	12.64	20.2	50.6	1024	3.1
5	19.02	32.9	95.1	3125	9.6
6	26.55	48.8	159.2	7776	23.9
7	35.20	68.2	246.4	16807	51.6
8	44.94	91.1	359.5	32768	100.6
9	55.75	117.1	501.9	59049	181.3
10	67.61	147.9	676.1	100000	307.1
11	80.49	181.9	885.4	161051	494.5
12	94.38	219.7	1132.6	248832	764.1
13	109.27	261.3	1420.7	371293	1140.1
14	125.14	307.0	1752.0	537824	1651.5
15	141.99	356.6	2129.8	759375	2331.8
16	159.79	410.1	2556.6	1048576	3219.9
17	178.53	467.8	3035.0	1419857	4360.0
18	198.22	529.4	3568.0	1889568	5802.3
19	218.84	595.4	4158.2	2476099	7603.4
20	240.37	665.6	4807.5	3200000	9826.3
21	262.82	740.1	5519.2	4084101	12541.1
22	286.18	818.5	6296.2	5153632	15825.3
23	310.43	901.4	7139.9	6436343	19764.1
24	335.57	988.5	8053.7	7962624	24450.9
25	361.60	1080.4	9040.0	9765625	29987.3
26	388.51	1176.3	10101.2	11881376	36484.2
27	416.29	1276.6	11239.6	14348907	44061.3
28	444.94	1381.5	12458.3	17210368	52848.0
29	474.45	1490.8	13759.0	20511149	62983.7
30	504.82	1604.6	15144.4	24300000	74618.2

RESISTANCE OF SHIPS.

Table XII.

Factors for Reduction of Speed and Power to Standard Displacement of 10,000 Tons.

Displacement.	Factor for Speed.	Factor for Power.	Displacement.	Factor for Speed.	Factor for Power.	Displacement.	Factor for Speed.	Factor for Power.
100	2.155	215.450	5500	1.105	2.009	10800	.987	.914
200	1.919	95.970	5600	1.102	1.967	10900	.986	.904
300	1.794	59.798	5700	1.098	1.927	11000	.984	.895
400	1.710	42.750	5800	1.095	1.888	11100	.983	.885
500	1.654	33.079	5900	1.092	1.851	11200	.981	.876
600	1.598	26.638	6000	1.089	1.815	11300	.980	.867
700	1.558	22.313	6100	1.086	1.780	11400	.978	.858
800	1.523	19.043	6200	1.083	1.747	11500	.977	.850
900	1.494	16.598	6300	1.080	1.714	11600	.975	.841
1000	1.468	14.678	6400	1.077	1.683	11700	.974	.833
1100	1.445	13.133	6500	1.074	1.653	11800	.973	.824
1200	1.424	11.866	6600	1.072	1.624	11900	.971	.816
1300	1.405	10.808	6700	1.069	1.596	12000	.970	.808
1400	1.388	9.913	6800	1.066	1.568	12100	.969	.801
1500	1.372	9.146	6900	1.064	1.542	12200	.967	.793
1600	1.357	8.483	7000	1.061	1.516	12300	.966	.785
1700	1.344	7.903	7100	1.059	1.491	12400	.965	.778
1800	1.331	7.394	7200	1.056	1.467	12500	.963	.771
1900	1.319	6.942	7300	1.054	1.444	12600	.962	.764
2000	1.308	6.538	7400	1.051	1.421	12700	.961	.757
2100	1.297	6.177	7500	1.049	1.403	12800	.960	.750
2200	1.287	5.850	7600	1.047	1.377	12900	.958	.743
2300	1.278	5.555	7700	1.045	1.357	13000	.957	.736
2400	1.269	5.286	7800	1.042	1.336	13100	.956	.730
2500	1.260	5.040	7900	1.040	1.317	13200	.955	.723
2600	1.252	4.814	8000	1.038	1.297	13300	.954	.717
2700	1.244	4.607	8100	1.036	1.279	13400	.952	.711
2800	1.236	4.416	8200	1.034	1.261	13500	.951	.705
2900	1.229	4.238	8300	1.032	1.243	13600	.950	.699
3000	1.222	4.074	8400	1.030	1.226	13700	.949	.693
3100	1.216	3.921	8500	1.028	1.209	13800	.948	.687
3200	1.209	3.779	8600	1.026	1.192	13900	.946	.681
3300	1.203	3.645	8700	1.024	1.176	14000	.945	.675
3400	1.197	3.521	8800	1.022	1.161	14100	.944	.670
3500	1.191	3.403	8900	1.020	1.146	14200	.943	.664
3600	1.186	3.293	9000	1.018	1.131	14300	.942	.658
3700	1.180	3.190	9100	1.016	1.116	14400	.941	.653
3800	1.175	3.092	9200	1.014	1.102	14500	.940	.648
3900	1.170	3.000	9300	1.012	1.088	14600	.939	.643
4000	1.165	2.912	9400	1.010	1.075	14700	.938	.638
4100	1.160	2.830	9500	1.008	1.062	14800	.937	.633
4200	1.156	2.751	9600	1.007	1.049	14900	.936	.628
4300	1.151	2.677	9700	1.005	1.036	15000	.935	.623
4400	1.146	2.606	9800	1.003	1.024	15100	.934	.618
4500	1.142	2.539	9900	1.001	1.012	15200	.933	.613
4600	1.138	2.474	10000	1.000	1.000	15300	.932	.609
4700	1.134	2.413	10100	.998	.988	15400	.931	.604
4800	1.130	2.354	10200	.997	.977	15500	.930	.600
4900	1.126	2.298	10300	.995	.966	15600	.929	.595
5000	1.122	2.245	10400	.993	.955	15700	.928	.590
5100	1.118	2.194	10500	.992	.945	15800	.927	.586
5200	1.115	2.145	10600	.990	.934	15900	.926	.582
5300	1.111	2.098	10700	.989	.924	16000	.925	.578
5400	1.108	2.052						

TABLES. 231

Table XIII.

Factors for Reduction of Dimensions to Standard Displacement of 10,000 Tons.

Displacement.	Factor.	Displacement.	Factor.	Displacement.	Factor.	Displacement.	Factor.	Displacement.	Factor.
100	4.642	3300	1.447	6500	1.154	9700	1.010	12900	.919
200	3.684	3400	1.433	6600	1.149	9800	1.007	13000	.916
300	3.218	3500	1.419	6700	1.143	9900	1.003	13100	.914
400	2.924	3600	1.406	6800	1.137	10000	1.000	13200	.912
500	2.714	3700	1.393	6900	1.132	10100	.997	13300	.909
600	2.554	3800	1.381	7000	1.126	10200	.993	13400	.907
700	2.426	3900	1.369	7100	1.121	10300	.990	13500	.905
800	2.321	4000	1.357	7200	1.116	10400	.987	13600	.903
900	2.231	4100	1.346	7300	1.110	10500	.984	13700	.900
1000	2.154	4200	1.335	7400	1.105	10600	.981	13800	.898
1100	2.087	4300	1.325	7500	1.100	10700	.978	13900	.896
1200	2.027	4400	1.315	7600	1.096	10800	.975	14000	.894
1300	1.983	4500	1.305	7700	1.091	10900	.972	14100	.892
1400	1.926	4600	1.295	7800	1.086	11000	.969	14200	.890
1500	1.882	4700	1.286	7900	1.082	11100	.966	14300	.888
1600	1.842	4800	1.277	8000	1.077	11200	.963	14400	.886
1700	1.805	4900	1.268	8100	1.073	11300	.960	14500	.883
1800	1.771	5000	1.260	8200	1.068	11400	.957	14600	.881
1900	1.739	5100	1.252	8300	1.064	11500	.955	14700	.879
2000	1.710	5200	1.244	8400	1.060	11600	.952	14800	.877
2100	1.682	5300	1.236	8500	1.056	11700	.949	14900	.876
2200	1.656	5400	1.228	8600	1.052	11800	.946	15000	.874
2300	1.632	5500	1.221	8700	1.048	11900	.944	15100	.872
2400	1.609	5600	1.213	8800	1.044	12000	.941	15200	.870
2500	1.587	5700	1.206	8900	1.039	12100	.938	15300	.868
2600	1.567	5800	1.199	9000	1.036	12200	.936	15400	.866
2700	1.547	5900	1.192	9100	1.032	12300	.933	15500	.864
2800	1.528	6000	1.186	9200	1.028	12400	.931	15600	.862
2900	1.511	6100	1.179	9300	1.025	12500	.928	15700	.860
3000	1.494	6200	1.173	9400	1.021	12600	.926	15800	.859
3100	1.477	6300	1.167	9500	1.017	12700	.923	15900	.857
3200	1.462	6400	1.160	9600	1.013	12800	.921	16000	.855

Table XIV.
Values of Slip Angle φ.

Slip. Diameter Ratio.	.05		.10		.15		.20		.25		.30		.35		.40	
	Deg.	Min.	Deg.	Min.	Deg.	Min.	Deg.	Min.	Deg.	Min.	Deg.	Min.	Deg.	Min.	Deg.	Min.
.1	0	51	1	48	2	51	4	00	5	17	6	44	8	21	10	12
.2	1	21	2	47	4	20	6	00	7	49	9	46	11	53	14	11
.3	1	28	3	01	4	39	6	22	8	11	10	06	12	06	14	13
.4	1	25	2	54	4	26	6	02	7	41	9	24	11	10	12	59
.5	1	19	2	40	4	04	5	30	6	58	8	28	10	00	11	35
.6	1	12	2	25	3	41	4	57	6	15	7	34	8	56	10	18
.7	1	5	2	12	3	19	4	28	5	38	6	48	7	59	9	12
.8	1	00	2	00	3	01	4	03	5	05	6	08	7	12	8	16
.9	0	54	1	49	2	45	3	41	4	37	5	34	6	32	7	30
1.0	0	50	1	40	2	31	3	22	4	14	5	06	5	58	6	51

TABLE XV.

Factors by which Old Pitch must be multiplied to obtain New Pitch after Blade is twisted.

Twist of Blade.	1°		2°		3°		4°		5°		6°	
	Factor for		Factor for		Factor for		Factor for		Factor for		Factor for	
Diameter Ratio.	Increase.	Decrease.	Increase.	Decrease.	Increase.	Decrease.	Increase.	Decrease.	Increase.	Decrease.	Increase.	Decrease.
.1	1.065	.943	1.138	.891	1.221	.843	1.315	.800	1.425	.761	1.553	.725
.2	1.040	.963	1.083	.927	1.127	.893	1.175	.861	1.226	.830	1.281	.800
.3	1.036	.966	1.073	.933	1.111	.901	1.152	.870	1.194	.840	1.237	.811
.4	1.037	.965	1.074	.930	1.112	.897	1.152	.864	1.193	.832	1.236	.801
.5	1.039	.962	1.079	.925	1.120	.886	1.162	.852	1.205	.817	1.249	.783
.6	1.043	.958	1.086	.918	1.130	.877	1.176	.837	1.222	.798	1.269	.760
.7	1.047	.954	1.094	.909	1.142	.864	1.192	.820	1.242	.777	1.293	.734
.8	1.051	.950	1.103	.900	1.156	.851	1.210	.802	1.264	.754	1.320	.707
.9	1.056	.945	1.113	.891	1.170	.837	1.229	.783	1.288	.730	1.348	.678
1.0	1.060	.940	1.122	.880	1.184	.821	1.247	.763	1.311	.705	1.376	.648

FORMULÆ.

Let p denote pitch before twisting; p', pitch after twisting.
Let θ denote pitch angle before twisting; θ', pitch angle after twisting.
Let y denote diameter ratio before twisting.
Let ϕ' denote angle of twist.   Then $\tan \theta' = \dfrac{p'}{p} \tan \theta.$ $\theta \pm \phi' = \theta'.$

TABLE XVI.
Functions for Characteristics for Standard Blade.

X_f for X_c

Diameter Ratio	.00	.01	.02	.03	.04	.05	.06	.07	.08	.09
.4	.170	.176	.183	.189	.195	.202	.208	.215	.222	.229
.5	.236	.243	.250	.257	.264	.272	.279	.286	.294	.301
.6	.309	.316	.323	.331	.338	.345	.353	.361	.368	.376
.7	.383	.391	.398	.406	.414	.422	.430	.438	.445	.453
.8	.461	.469	.471	.479	.492	.500	.508	.516	.523	.531
.9	.539	.546	.555	.562	.570	.578	.586	.593	.601	.609

Y_f for Y_c

Diameter Ratio	.00	.01	.02	.03	.04	.05	.06	.07	.08	.09
.4	.524	.527	.530	.533	.537	.540	.543	.547	.550	.553
.5	.557	.560	.564	.568	.572	.576	.580	.585	.589	.593
.6	.598	.603	.608	.613	.618	.623	.628	.634	.640	.645
.7	.652	.658	.664	.670	.676	.682	.689	.695	.702	.708
.8	.714	.721	.727	.733	.740	.746	.752	.759	.765	.772
.9	.778	.784	.791	.797	.804	.810	.817	.823	.830	.837

Z_f for Z_c

Diameter Ratio	.00	.01	.02	.03	.04	.05	.06	.07	.08	.09
.4	0.48	0.50	0.51	0.52	0.54	0.55	0.57	0.59	0.61	0.63
.5	0.65	0.68	0.70	0.73	0.75	0.78	0.81	0.84	0.87	0.91
.6	0.94	0.98	1.01	1.05	1.10	1.14	1.19	1.24	1.29	1.34
.7	1.40	1.45	1.51	1.57	1.63	1.69	1.76	1.83	1.90	1.97
.8	2.04	2.11	2.19	2.27	2.34	2.43	2.51	2.60	2.68	2.77
.9	2.87	2.97	3.06	3.15	3.25	3.36	3.46	3.57	3.68	3.78

This table is filled out for each hundredth of diameter ratio.
Full tenths are indicated in the left-hand column and hundredths in the top line.

www.ingramcontent.com/pod-product-compliance
Lightning Source LLC
Chambersburg PA
CBHW021702230426
43668CB00008B/703